AF553689

ENCYCLOPAEDIA OF ENDOCRINOLOGY-III

REPTILIAN ENDOCRINOLOGY

By

Manju Yadav

Lecturer

Department of Zoology

M.M.H. College

Ghaziabad (U.P.)

(India)

DISCOVERY PUBLISHING HOUSE PVT. LTD.

NEW DELHI-110 002

First Published-2008

ISBN 978-81-8356-273-7

Published by

DISCOVERY PUBLISHING HOUSE PVT. LTD.
4831/24, Ansari Road, Prahlad Street,
Darya Ganj, New Delhi-110002 (India)
Phone: 23279245 • Fax: 91-11-23253475
E-mail: dphbooks@rediffmail.com
dphtemp@indiatimes.com

Printed at:

Sachin Printers, Delhi

Preface

The present title **Reptilian Endocrinology** provides a definitive and comprehensive review of all aspects relating to hormones in different animals including wild and domestic species. It discusses the intimate physiology of the endocrine system itself and describes the role of hormones in the processes of nutrition, osmoregulation, colour change, calcium metabolism and reproduction. Subject matter has been presented in a readable way emphasizing the differences between species from a functional aspects. It not only contains a wealth of information concerning the endocrinology of the various species but also contains more than enough basic information concerning the biosynthesis, metabolism and mode of action of the hormones for the book to be read and understood without referring to a text book of basic endocrinology. This book has been written primarily for use as a text book by under-graduate, as well as graduate students.

Efforts have also been made to make the presentation lucid and accessible for beginning students, emphasizing major concepts while in corporating the latest experimental findings and theories throughout. The aim is to enthuse the readers with this active and exciting area of research and to lay a solid foundation on which further study of its various facts may be based.

Though, the author has taken special care to present a current account, yet he is fully aware about limitations and the readers may come across the mistakes of various types. For all types of mistakes author extends his due apology.

There can be no claim to originality except in the manner of treatment and much of the information has been obtained from the books and scientific journals available in the different libraries.

The author expresses his thanks to his friends and colleagues whose continue inspirations have initiated him to bring out this book.

The author expresses his gratitude to Mr. Wasan and staff of M/s Discovery Publishing House for their whole hearted co-operation in the publication of this book.

Author

Contents

1

INTRODUCTION

Vertebrate animals, whether warm or cold blooded, have adapted to a great variety of aquatic or terrestrial conditions. They have achieved this tremendous diversity of habitat with a seemingly large variety of anatomical and physiological adaptations. Although it is true that by virtue of special modifications, certain vertebrates have been more successful than others, it is most striking that all the various modifications have been made with the same machinery used in a fundamentally similar fashion. A detailed study of vertebrate physiology is indispensable to understand the vertebrate world and man himself.

Reptiles are cold blooded vertebrates living and reproducing under the conditions imposed by a land environment. Many fishes have enzymatic adaptations freeing them from the barriers of cold. Birds and mammals have enjoyed the great liberation provided by warm bloodedness.

The warmer Mesozoic Era was favourable to the great development and diversification of the reptiles, but the colder modern earth has imposed severe limitations on the land-living and breeding cold blooded vertebrates.

Our present task is to summarize the information concerning the reproductive physiology of some 6000 species of turtles, alligators, lizards, and snakes. Although this is a fairly large and diverse group of animals, adequate data on reproduction is available for less than 1 per cent of the known number of species. This discussion will cover the morphology and physiology of the testis, male accessory organs, secondary sexual characteristics, the ovary, female accessories, and, finally, the factors related to oviparity, ovoviviparity, and viviparity.

Testis

The reptilian testis is anatomically similar to that of birds and mammals. Essentially it is an ovoid body suspended in the posterior part of the body cavity by a mesentery. It is made up of convoluted tubules held together by a relatively thin connective tissue envelope. Lying between the tubules are small amounts of connective tissue and interstitial cells. The tubular elements are Sertoli cells, spermatogonia, and the products of spermatogonial divisions and spermatidal transformations. Spermatogenesis and spermiogenesis are cyclical and climatically related phenomena.

The testicular cycle has been described for turtles, lizards, and snakes, but little is known of the alligator cycle. In turtles, spermatogenesis occurs primarily in the late spring and summer; spermiogenesis takes place in the fall. The mature sperm usually pass the winter hibernation period in the sex accessories.

Although the testicular cycle of turtles is like that of the amphibia, the cycle of lizards is similar to that of birds. Although there is considerable variation among lizards regarding the exact sequence of events, the most common type of testicular cycle is one wherein spermatogenesis occurs in late summer and fall, and occasionally during the period of winter hibernation. Spermiogenesis may begin in the fall or winter, but is largely a spring phenomenon. Passage of mature sperm into the epididymides, and copulation, usually occur in the spring.

Hamlett (1952) and Fox (1958) pointed out that in the more southern living American chameleon, *Anolis carolinensis*, the testicular cycle is unusual, in that the testis and sex accessories are maintained in a fully mature condition throughout the summer (from April until August). This situation makes more apparent the fact that our knowledge of reptilian reproductive cycles is based almost exclusively on north and south temperate living animals, and that very little is known of tropical species.

Among the snakes, the most usual type of testicular cycle is similar to the lizard. In most European snakes, the sperm are matured and shed in the spring, the testis is quiescent during the summer, and a new cycle is reinstituted in the fall. In the North American garter snakes, the testes are maximally developed during the summer and are regressed in the winter. Spermatogenesis and spermiogenesis occur in the spring.

Both males and females of the northern races of certain snakes and lizards follow a biennial schedule of reproduction. This has been described in *Crotalus viridis*, *Vipera berus*, and in *Heloderma suspectum*. A biennial cycle probably occurs also in mountain living individuals of the viviparous, *Xantusia vigilis*.

Interstitial Cell

The Leydig cells occur as isolated groups of a few cells or as small compact masses. In the majority of species studied, a definite increase in interstitial cell activity is coincident with the actual time of breeding. Increased interstitial cell activity is indicated by an increase in cell number, cell volume, nuclear volume, and by a loss of lipids, cholesterol, or steroid-like compounds.

At the time of maximum interstitial cell activity, there is enlargement and increased secretory activity of the epididymides, the vas deferens are sperm-filled, the hemipenes are enlarged, and other secondary sexual characteristics are developed.

A few species have been described in which Leydig cells were present but failed to show seasonal morphologic variations. In the lizard, *Sceloporus undulatus*, they may be completely absent. A remarkable feature was shown in a photograph of the testis of the Algerian water turtle, *Emys leprosa*, by Combescot, 1954. In this testis, a ring of hypertrophied lipid containing connective tissue cells was seen surrounding the evacuated tubules. The same condition prevails in most urodele amphibians and in certain fishes. Thus, we see a parallel type of evolution in certain fishes, urodele amphibians, and in a turtle, where a modified lipid-containing connective tissue ring represents the interstitial tissue. In most fishes, anuran amphibians, most reptiles, and all birds and mammals, scattered or clumped intertubular connective tissue-derived cells represent the other and most common type of interstitial cell elements.

The Sertoli cells in the reptilian testis not only play the usual nutritive role in the development of the mature sperm, but also act as phagocytic elements after the shedding of the sperm, as well as possibly salvaging lipids and steroids important to the endocrine economy of the animal. Although the morphologic evidence points to the importance of the interstitial cell as the site of production of male sex hormone, the possible role of both the Sertoli and seminal elements should not be overlooked.

Sexual Accessory Organs

In the male reptile these are the epididymides, the vas deferens, and modified elements of the metanephros, the so-called "renal sex segments." In all reptiles so far examined, the epididymal epithelium becomes hypertrophic and secretory at the time mature sperm are, formed in the testis. As long as the sperm are retained within the lumen, the epididymal epithelium remains actively secretory. In lizards, where the sperm are matured in the spring and shed into the epididymis shortly before copulation, the epididymides need only be enlarged for a short time. In turtles and certain snakes, the epididymis retains the sperm throughout the winter hiberation, and the secretory lining persists during this entire period.

The ductus deferens is not secretory in snakes, but may be in lizards. Unfortunately, little is known concerning the lower portion of the genital tract in male lizards.

A portion of the nephron in the metanephros of male lizards and snakes is a prominent sexual accessory organ, the "renal sex segment." The renal sex segment does not appear in females, or in turtles or alligators. Some workers have described the pre-terminal portion of the nephron as the enlarged segment, whereas others have found the entire terminal nephric segment, the collecting ducts, and the ureters to be involved. Some of these differences may be species variations, but some may be merely a matter of anatomical interpretation.

Regardless of the precise part of the nephron involved at the time of maximum testicular activity (presumably when there is abundant male sex hormone present), the epithelial lining of the terminal portion of the nephron undergoes marked hypertrophy and the cytoplasm of the epithelium is charged with eosinophilic albuminoid material. These segments are large enough to be discernible with the naked eye. It has been suggested that the secretion produced by these cells acts as a nutritive medium for the sperm while they are in the cloaca.

The development and duration of the renal sex segment, as in the epididymis, corresponds to the type of testicular cycle. In lizards, the renal sex segments develop only during the spring breeding season and are regressed during the remainder of the year. In snakes containing overwintering epididymal sperm, the renal sex segments are enlarged throughout the year.

Secondary Sexual Characteristics

The development of secondary sexual characteristics in reptiles is not nearly as great as in fishes and birds. Among the reptiles, the lizards are the best endowed, whereas snakes and turtles exhibit little in the way of sexual dimorphism. In many lizard species, such features as body size, post-anal swelling due to the hemipenes, femoral pores, dorsal crests, gular folds or pouches, and colour differences distinguish males from females. Some turtles show differences in the concavity of the plastron, the colour of the iris, and in the length of the tail and certain of the hind toe claws. Snakes show very little in the way of sexual dimorphism, this being limited to variation in body and tail length, differences in scale counts, and the presence of penile sacs.

Certain of the secondary sexual characteristics, particularly in the lizards, develop and regress coincidentally with the waxing and waning of the testis. These characteristics are the hemipenes, the femoral glands, gular folds and pouches, and certain types of colouration.

Relation of Male Sex Hormone to the Growth and Development of the Sexual Accessories and Secondary Sexual Characteristics

Castration of juvenile or adult reptiles will prevent the growth and development of the epididymides, the renal sex segments, and the male secondary sex characteristics. Restoration of male sex hormone by testicular grafts or the administration of male sex hormone will restore the growth and development of these structures. Male hormones given to juveniles or adults with seasonally quiescent reproductive structures may induce precocious growth and development of the accessories and secondary characteristics, and testicular spermatogenesis, but may effect a reduction of the testis volume.

The administration of female sex hormone to male reptiles may either stimulate or inhibit the development of the sex accessories. The administration of male sex hormone to female reptiles may stimulate the oviducts or bring about the development of the renal sex segments in female lizards and snakes. Dantschakoff (1937) and Dantschakoff and Kinderis (1938) made the very important observation that, in reptiles and birds, the male is the homochromic sex (male = XX constitution), and that the sex hormone of the heterochromic female is dominant. Thus, female sex hormones administered to male reptiles stimulated the development of the plastic mullerian duct, inhibited the penis and suppressed the medullary portion of the testis.

Relationship of the Pituitary Gland to the Testis

The pituitary gland of reptiles has been studied in recent years by Miller (1948b), Cieslak (1945), and Hartman (1944). It is generally agreed that an increase in the secretory activity of the basophil cell of the pars anterior coincides with the time of greatest gonadal activity. These studies were made before improvements in pituitary staining were developed by Romeis and Gomori and utilized so effectively by

Halmi (1950) and Purves and Griesbach (1951). Very recently re-examination of the pituitary gland of the viviparous lizard, *Xantusia vigilis*, with the periodic acid Schiff and aldehyde fuchsin stains unquestionably revealed that the old aniline-blue-staining basophil is the same as the delta basophil of modern terminology. Thus, it may be the homologue of the supposed follicle stimulating hormone-producing basophil of the mammalian pituitary gland. With the Heidenhain Azan or Mallory Triple stains, the anterior pituitary of reptiles shows two types of acidophil cells. Use of the newer techniques, however, has shown that one of these so-called acidophils contains granules which are both periodic acid Schiff and aldehyde fuchsin positive and therefore may be homologous with the thyrotrophic beta basophil of the mammal. Significantly, by the application of these techniques, the anterior pituitary gland of fishes, amphibians, reptiles, birds, and mammals appear to contain a morphologically homologous series of cell types. Whether or not these cell types are functionally homologous remains to be determined.

The effects of hypophysectomy in reptiles indicates that gonadal growth and development, and indirectly the growth and maintenance of the sex accessories and secondary characteristics, is dependent upon anterior pituitary hormones. As a rule the effect of these materials was stimulation of the testis, the accessories, and the secondary sexual characteristics.

Bartholomew (1950) has reviewed the literature concerning the effect of light and temperature on gonadal development in reptiles. Generally, an increase in the amount of light, along with an appropriate increase in temperature, will stimulate the gonads. However, Bartholomew suggests that light may only reinforce an internally controlled reproductive rhythm rather than act as the primary controller of the reproductive cycle.

Relation of Other Endocrine Glands to Reproduction

The thyroid gland shows definite seasonal variations in activity in reptiles. In the male, greatest thyroid activity is observed during the period of breeding (usually at the time of testicular maturation and copulation). It is also true that in some more southern living reptiles, which are able to remain active throughout the winter period, the thyroid gland is concomitantly active.

It is difficult to assess the exact relation of the thyroid gland to reproduction, since the breeding period is associated also with intense feeding and fighting and greatest overall activity.

Studies on the interrenal tissue of reptiles by Miller (1955) and Fox (1952) show that there is increased activity of these glands during the final stages of gonadal maturation, development of the sexual accessories, and the period of copulation.

In general, one may conclude that the endocrine regulation of reproduction in male reptiles is much the same as in mammals.

Ovary

The ovaries of reptiles are similar to those of amphibia and birds. They consist of a thin stromal wall in which the developing ova are clearly visible, a variously placed patch of germinal epithelium, and a lymph filled central cavity. At certain times one may find corpora atretrica or corpora lutea.

The ova of reptiles living in warmer climates may grow to maturity within a few months' time. In most strictly north temperate species development of ripe ova may require two, three, or even four years.

Several types of ovarian cycles are found in reptiles, it is not yet possible to correlate the type of cycle with taxonomic, geographic, or climatic conditions. The following types of ovarian cycles have been described in reptiles: (1) A large amount of yolk is deposited shortly before ovulation, but is subsequent to a long slow initial growth of the ova (type found in *Phrynosoma*, *Hemidactylus*, and *Xantusia*); (2) Yolk deposition occurs gradually during most of the year preceding ovulation (type found in *Lacerta agilis* and other European lizards); (3) Yolk deposition occurs shortly after ovulation and mature ova remain in the ovary throughout the winter (type found in *Sceloporus graciosus*); (4) Two sets of ova are produced and ovulated annually, each set is formed directly after ovulation of the preceding set (type found in *Amphibolurus*); (5) Nonseasonal breeders which may reproduce at any time of the year (type reported for certain Javanese snakes and for the New Hebrides lizard, *Lygosoma*).

Follicular atresia is a common occurrence in reptiles. In the process of atresia, the theca cells undergo marked hyperplasia and phagocytize the yolk. Since follicular atresia is a very active process in the maturing ovary, the secretory-appearing cells of the theca must be considered as a possible source of estrogens. These structures are similar to the corpora atretica found in the ovaries of certain fish and amphibia.

Corpus Luteum

Apparently all reptiles, whether oviparous or viviparous, develop a true post-ovulatory corpus luteum. This structure is produced by the hyperplasia of both the granulosa epithelium and the surrounding thecal elements. The degree of involvement of each of these is variable and bears no relation to taxonomic affinities or to the degree of viviparity. The most common type of corpus is composed of a central mass of enlarged, lipid-containing granulosa-derived luteal cells, surrounded by a connective tissue-derived thecal envelope, which may or may not send penetrating strands into the luteal tissue.

There is a definite correlation between the longevity of the corpus luteum and the egg-laying or retaining habit of the species. Oviparous reptiles have a corpus luteum regressing shortly after egg-laying. In ovoviviparous and viviparous species the corpus luteum remains developed for approximately three-quarters of the time of gestation. The average gestation time of most north and south temperate living reptiles is three months. In these species the corpus luteum remains developed for approximately two months and begins to regress during the latter third of gestation.

The specific function of the corpus luteum, however, remains an enigma. The earlier studies of Clausen (1940) and Fraenkel et al. (1940) indicated the corpus luteum was essential for maintaining pregnancy in viviparous snakes. In their experiments, deluteinization during early pregnancy was followed by abortion or resorbtion of embryos. The more recent experiments of Bragdon (1951) on the viviparous garter snake, and of Panigel (1956) on the ovoviparous lizard, *Zootoca*, indicate that the corpus luteum is not essential for the maintenance of gestation. As Bragdon suggests, it will be necessary to study more ovoviviparous species, and particularly the more completely viviparous species before drawing definite conclusions. One might suggest this study include the Australian snakes of the genus *Denisonia*, the lizards of the genus *Lygosoma*, and the old world skink, *Chalcides ocellatus*.

Relationship of Female Sex Hormones to the Growth and Development of the Female Sex Accessories

In general, ovariectomy results in regression of the oviducts, whereas administration of estrogens stimulates the growth and development of these structures. It is interesting that testosterone given in proper amounts will also stimulate the oviduct, just as estrogens may sometimes stimulate the epididymides. Progesterone, alone, or in combination with estrogens, has not been studied adequately. One report by Panigel (1956) shows that although progesterone will stimulate the oviduct, it is less effective than either estrogen or testosterone. Panigel believes further that estrogens may act most effectively on the glandular portion of the oviduct, and progesterone on the muscular wall.

Significantly, a progesterone-like substance has been demonstrated in the extracts of ovaries containing corpora lutea, and in the plasma of pregnant viviparous snakes. Furthermore, in certain viviparous reptiles, if only one oviduct is occupied by a developing embryo, the other will remain in an enlarged state throughout the entire period of gestation indicating the presence of ovarian or placental hormones.

The specific role of ovarian hormones, whatever their nature or source, remains undetermined in the reptiles.

Relation of the Pituitary Gland to the Ovary and Gestation

Hypophysectomy performed before ovulation produces regression of the developing ovary and oviducts. Hypophysectomy performed during gestation, however, will not usually lead to abortion, but will interfere with parturition. The mechanism of the latter is unknown. Laboratory conditions alone are inimical to normal parturition in viviparous lizards.

The administration of anterior pituitary extracts is effective in stimulating both ovaries and oviducts.

As in the male, light is capable of precociously stimulating the ovary and oviducts.

Cytologic studies of the reptilian pituitary gland indicate that the delta-basophil is the most active cell during yolk deposition and ovulation. During gestation in the viviparous *Xantusia vigilis*, the beta-basophil (possibly the thyrotrophin producing cell) is the most actively secretory type.

Relation of Other Endocrine Glands to the Ovary and Gestation

Both the thyroid and interrenal glands show heightened secretory activity at the time of final yolk deposition, ovulation, and the early phases of gestation.

Oviparity, ovoviviparity, and viviparity

Although oviparity is the most common method of reproduction in reptiles and is the only type found in alligators and turtles, many lizards and snakes have evolved ovoviviparous and viviparous conditions. The author would like to propose that the term ovoviviparity be used for the condition in which a shell encloses a developing embryo habitually retained in utero for a certain length of time, and viviparity, a condition in which no shell, but a thin shell membrane may or may not enclose the developing embryo. Embryos enclosed by a shell membrane are included in the viviparous group, since the relation between oviduct and extraembryonic membranes may be as advanced as in many species having no interposing shell membrane.

In ovoviviparous and viviparous reptiles there are differences in the extraembryonic membranes and the relation of these to the oviducal wall. Notably, in all cases, a yolk sac placenta is first formed, but usually is succeeded by a chorioallantoic type of placenta of varying magnitude and complexity. In only one reptilian species so far described, *Gongylus ocellatus* does the yolk sac play a major role throughout gestation. In all other forms, the allantois assumes the major role and in combination with the chorion develops into a chorioallantoic placenta. The degree of development of the chorioallantois and the nature of the relationship of the oviducal wall to the placental membrane ranges from a rather loose apposition of flat epithelial surfaces to the formation of complex interdigitating villal folds. In more advanced grades of viviparity, particular portions of the placenta are specialized for the better apposition of the maternal and fetal circulations.

The glandular tissue of the oviducal wall in some cases secretes substances which may be absorbed by either the yolk sac or the chorioallantoic membranes, an embryotrophic type of nutrition.

Certainly in most instances, there is at least exchange of gases and water, and probably even of nutritive materials across the oviducal epithelium-yolk-sac or oviducal epithelium-chorioallantoic membranes. Interestingly, the weight of a fully developed newborn animal is approximately twice that of the freshly ovulated ovum, and, hence, half the weight of the developed young reptiles must be materials gained from the maternal circulation. It is also true, however, that both lizard and turtle eggs developing outside the mother's body gain considerable weight by absorption of water through the egg shells.

Placental transmission apparently occurs in some viviparous species to the extent that a considerable portion of the original yolk is unused at the time of birth. This material is retracted within the abdominal cavity and provides the young animal with food for several months (in *Vipera berus*, and in *Xantusia vigilis*).

The studies of Clarke et al. (1955), Clarke (1953), and of Bellairs et al. (1955), indicate that amino acids are probably transmitted from the maternal to the fetal circulation. Clarke (1953) has interestingly demonstrated that urea is the major product of early embryonic nitrogen excretion in the black snake, and that it may be transmitted to the mother. One might speculate on the role of retention of urea excretion in the early phases of embryonic development in reptiles in relation to the ability of these forms to become viviparous. A very important observation in relation to ovoviviparity and viviparity in reptiles is that those species living at the highest latitudes both north and south and at the highest altitudes are almost always ovoviviparous or viviparous.

There is no doubt that ovoviviparity and viviparity have developed independently many different times in a wide variety of reptilian families. It is particularly significant that although this phenomenon has occurred independently many times, the mechanism of its accomplishment has been remarkably similar.

Differences in the endocrine regulation of oviparous, ovoviviparous, and viviparous methods of reproduction have yet to be determined.

2

PITUITARY GLAND

The important role of the pituitary as an endocrine gland became progressively more obvious at the end of the last and beginning of the century. More recently the discovery of the adenohypophyseal mechanism controlling the hypothalamic neurosecretory products has caused a renewal of interest in the whole of that region, and now numerous publications appear on it each year. However, as in many other studies, our knowledge of comparative morphology and physiology lags behind that concerning the laboratory mammals, and the reptiles are, of all the vertebrates, those which have been studied least.

Since Rathke's publications it has been known that the morphology of the pituitary in snakes is basically similar to that of the other tetrapod vertebrates. Numerous later works, of which the oldest generally deal with embryology, have discussed the pituitary in other snakes, in lizards, in *Sphenodon*, in turtles, and in crocodilians.

Particular attention should be called to the important works on embryology or comparative morphology dealing with representatives of several orders of reptiles. Finally, I have recently described numerous additional species. Table 2.1 indicates that the morphology of the reptilian pituitary is known in representatives of all the suborders and many families. It is therefore possible and desirable to give a general description rather than to analyze in sequence the numerous publications which have been devoted to it.

The terminology used to designate the different parts of the hypophysis and hypothalamic-neurohypophyseal complex varies considerably with different authors. Table 2.2, which includes only terms used in comparative morphology, will help to clarify these.

Table 2.1. List giving the number of species in each family in which the morphology of the hypophysis is known.

Rhynchocephalia		Ophidia	
Sphenodontidae	1	*Typhlopoidea*	
Sauria		Typhlopidae	3
Gekkota		*Leptotyphlopoidea*	
Gekkonidae	10	Leptotyphlopidae	1
Pygopodidae	2	*Booidea*	
Xantusiidae	2	Boidae	5
Iguania		*Colubroidea*	
Iguanidae	15	Colubridae	13
Agamidae	9	Elapidae	9
Chamaeleonidae	3	Hydrophiidae	3
Scincomorpha		Viperidae (including Crotalinae)	7
Scincidae	12		
Feyliniidae	1	Testudines	
Lacertidae	9	*Cryptodira*	
Cordylidae (including Gerrhosaurinae)	3	Testudinidae (including Emydinae)	6
Anguimorpha		*Pleurodira*	
Anguidae	3	Pelomedusidae	1
Anniellidae	1	Crocodilia	
Helodermatidae	2	Crocodylidae	4
Varanidae	2		
Amphisbaenia			
Amphisbaenidae	1		
Trogonophidae	1		

Hypothalamic-Neurohypophyseal Complex

Although the anatomy of the neural lobe of the hypophysis of reptiles has long been known, the idea of neurosecretion is much more recent, and their hypothalamic-neurohypophyseal relations have been the object of only a few studies. The neurosecretory phenomena in reptiles have most recently been dealt with by Gabe (1966).

Neurosecretory Perikarya

The neurosecretory perikarya of the hypothalamic-hypophyseal tract in reptiles are generally small, few in number, and gathered together

Table 2.2. Terminology used to designate the various parts of the hypothalamic-hypophyseal complex

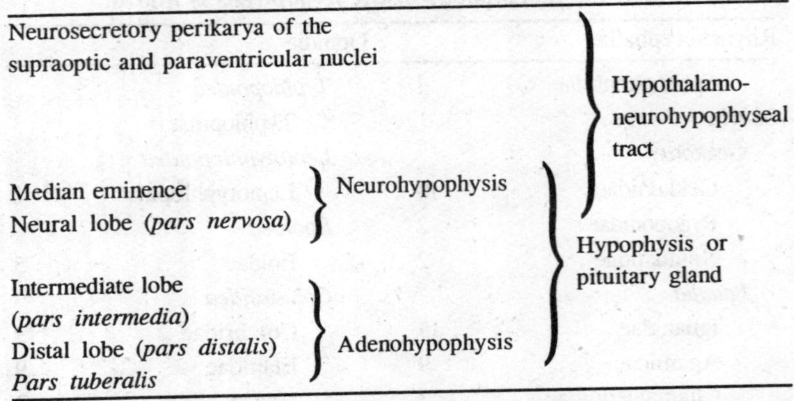

Part			
Neurosecretory perikarya of the supraoptic and paraventricular nuclei			Hypothalamo-neurohypophyseal tract
Median eminence	Neurohypophysis	Hypophysis or pituitary gland	Hypothalamo-neurohypophyseal tract
Neural lobe (*pars nervosa*)	Neurohypophysis	Hypophysis or pituitary gland	Hypothalamo-neurohypophyseal tract
Intermediate lobe (*pars intermedia*)	Adenohypophysis	Hypophysis or pituitary gland	
Distal lobe (*pars distalis*)	Adenohypophysis	Hypophysis or pituitary gland	
Pars tuberalis	Adenohypophysis	Hypophysis or pituitary gland	

in the supraoptic and paraventricular nuclei. The supraoptic nuclei form two distinct groups of cells lying dorsal to the optic tract at the point where it joins the diencephalon. The crescentic paraventricular nucleus is situated anterodorsal to the supraoptic nuclei. In most reptiles, the supraoptic nuclei are small and rich in neurosecretory cells; the paraventricular nucleus is often more extensive, but, except among the Testudines, has few neurosecretory elements.

The size and appearance of the neurosecretory perikarya vary greatly, perhaps from one species to another, but chiefly according to the functional phase of their secretory cycle. The only nearly constant characteristic is the presence of a large, spherical, basal nucleus, averaging 6 μ in diameter, with a very low chromatin content and a distinct nucleolus. The perikarya may be reduced to a thin border of cytoplasm in which some granules can be distinguished. More generally, they are ovoid and reach a size of 13 by 6 μ. Little or very little secretion is discernible with chromic hematoxylin, paraldehyde fuchsin, or alcian blue after permanganate oxidation. There are one or more small peripheral vacuoles. Finally, in some individuals, most neurosecretory cells are of an average size and completely filled with granules of variable size. Of course all the neurosecretory perikarya from one individual are never at the same stage, but usually the majority fall into one of the patterns which have just been described.

The descending axons from the neurosecretory perikarya of the paraventricular and supraoptic nuclei converge towards the floor of the third ventricle and attach themselves along the ventral wall of the infundibular recess. Their pathways are marked by an irregular string of small spherical granulations, occasionally intersected by pools of colloidal material; these are often numerous among the perikarya.

Median Eminence

Just anterior or dorsal to the anterior end of the pars distalis, the ventral wall of the infundibular recess forms a distinct swelling, the median eminence. A transverse section shows, from dorsal to ventral: (1) a layer of ependymal cells; (2) a layer mainly composed of pituicytes, but containing a certain number of horizontal neurosecretory fibres; (3) a thicker layer than the preceding ones, formed mostly by rectilinear and vertical neurosecretory fibres, as well as by some extensions from the basal poles of the ependymal cells, and also by some pituicytes; and (4) an external limiting layer of glia. Above this there is a mass containing much connective tissue and many capillaries. In most reptiles, these capillaries lift the limiting glia locally, without crossing it, and are thus in direct contact with the original fibres of the neurosecretory perikarya. In *Sphenodon punctatus*, on the contrary, the neurosecretory fibres cross over the ventral limiting glia and pass in front of the primary capillary network of the median eminence.

In the Rhynchocephalia, Testudines, and Crocodilia, the base of the median, eminence and a fairly large portion of the infundibular floor are overlapped by the cellular cords of the pars tuberalis. These are vestigial or absent in the Squamata.

In the Colubroidea, the median eminence is joined to the anterior end of the pars distalis by a distinct connecting tract, sometimes called the pars terminalis. In most other squamates and in *Sphenodon* the median eminence is situated nearer the neural lobe, immediately dorsal to the anterior third of the pars distalis. Its contact with the latter is generally fairly diffuse. In the Testudines and Crocodilia the position is the same as in the Sauria, but a thick pars tuberalis, especially in turtles, forms a large connecting area containing the vessels of the hypophyseal portal system.

In the Leptotyphlopidae, the median eminence is clearly hypertrophic and larger than the pars nervosa. As in all the limbless burrowing Squamata, it is closely applied to the diencephalic floor.

Neural Lobe

The spherical or ovoid, more or less lobulated pars nervosa forms a simple extended cul-de-sac of the infundibular recess, situated dorsal to the pars distalis in the majority of reptiles. However, the neural lobe is dorsolateral to the distal lobe in snakes, and in a distinctly lateral or posterior position in certain limbless, burrowing squamates.

In the Rhynchocephalia, Sauria, Testudines, and Crocodilia, the infundibular recess penetrates deeply into the pars nervosa, where it often forms very extensive hollow lobules laterally. In snakes the neural lobe is massive, solid, and incompletely divided into lobules by connective tissue septa. In the first case, the structure of the pars nervosa is reminiscent of the median eminence; two layers can be distinguished between the opening in the infundibular recess and the external limiting membrane, first a layer of ependymal cells and then a fairly thick mass formed by a combination of pituicytes, fibres arising from the neurosecretory perikarya, and extensions from the basal poles of the ependymal cells. The pituicytes are more abundant in the central region and the neurosecretory fibres in the peripheral region. In snakes the ependymal cells are present only at the anterior extremity of the pars nervosa which is composed almost entirely of a large mass of clustered neurosecretory fibres with occasional pituicytes. The neural lobe tends to become massive in the limbless burrowing lizards and amphisbaenians as in the Ophidia. The infundibular recess penetrates the lobe more deeply, but its opening is generally very flattened, and the principal mass of the organ is formed by a thick layer of clustered neurosecretory fibres.

In the neural lobe, the neurosecretory products are visible as granules and particles of a colloidal appearance. Their abundance varies greatly among individuals, but they are generally much more distinct in lizards than in snakes.

The ultrastructural characters of the different parts of the hypothalamo-neurohypophyseal tract have been studied in only a few reptiles: *Natrix natrix*, *Gekko japonicus*, and *Clemmys japonica*. They do not appear to have any notable differences from those in other amniotes.

ADENOHYPOPHYSIS

The adenohypophysis is not an anatomical entity; the intermediate lobe is closely associated with the neural lobe, and is joined to the distal lobe only by an often narrow isthmus. However, embryological studies show that the pars intermedia and the pars distalis have a common origin, and the histological structure confirms the relationship of these two lobes; both of them are composed of classic glandular cells.

Intermediate Lobe

The intermediate lobe, which may be thought of as composed of two layers of cells separated by the hypophyseal cleft, encloses the posteroventral half of the neural lobe, from which it is separated by a double membrane of connective tissue. The internal layer, which is in contact with the pars nervosa, is always well developed and is simple or pseudostratified. The external layer is fairly flat, even endothelial in appearance. In a narrow area at the posterior or posteroventral part of the intermediate lobe, the external layer continues without a break in continuity into the distal lobe, and some cells originally from the latter may invade it, particularly towards the ventral area. Numerous capillaries lie between the connective tissue membranes which separate the pars intermedia from the pars nervosa.

This basic arrangement is subject to numerous variations. In the Gekkonidae the intermediate lobe almost completely surrounds the neural lobe and extends forward to the connection of the pituitary stalk; on the other hand, in some Iguanidae and Lacertoidea, as well as in the Testudines, it surrounds only the posteroventral third of the pars nervosa. Finally, in larger species, the internal layer is more or less festooned with and sometimes composed of cellular cords of irregular size and shape. This phenomenon is particularly marked in the Varanidae, which form a transition to the second arrangement described below.

In chamaeleonids, iguanids of the genus *Anolis*, agamids, at least one lacertid, and one crocodilian, the intermediate lobe is distinctly hypertrophied compared with the rest of the hypophysis. In these cases the two cellular layers described above cannot be distinguished, and the entire organ is formed of fairly regular, thick cords; the reduced hypophyseal cleft is restricted to the posteroventral part of the lobe. There are a fair number of capillaries between the cellular cords of the pars intermedia. Anatomically, the intermediate lobe of *Crocodylus niloticus* is of this type, but the cellular cords are much thinner and more irregular.

Morphologically the intermediate lobe of snakes is fairly distinctive and constant. It encloses the posterior half of the neural lobe with a fairly thin layer of cells which sends digitations between the lobules of the latter; its principal mass, formed from irregular cellular cords, is situated between the pars nervosa and the pars distalis and communicates extensively with the latter. Numerous isolated cells of the intermediate lobe are found throughout the posterodorsal region of the distal lobe.

Finally, in fossorial reptiles the intermediate lobe is more or less atrophied. During the first stages of reduction in the Boidae, burrowing Viperidae, and Anniellidae there is a simple decrease in size, and the general form characteristic of the family or superfamily is retained. In the Feyliniidae and the Amphisbaenia the intermediate lobe is reduced to a thin strip joining the neural lobe to the posterior end of the distal lobe. In the Typhlopidae and Leptotyphlopidae, the pars intermedia has completely disappeared.

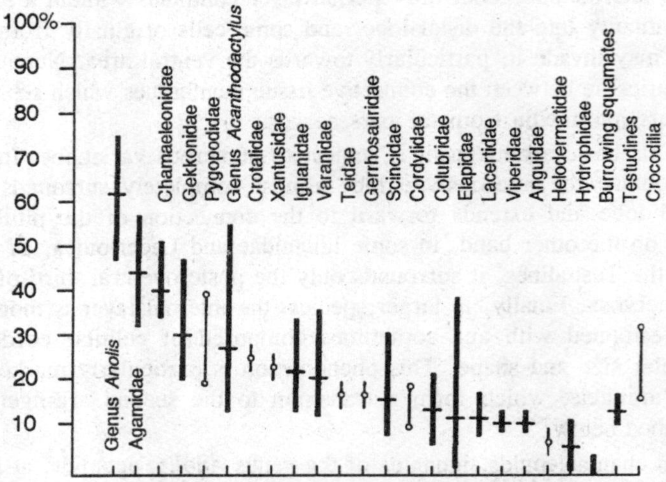

Fig. 2.1. Size of the intermediate lobe, expressed as a percentage of the entire hypophysis, in different families of reptiles.

When the internal layer is formed by a simple epithelium, the cells of the intermediate lobe are regularly prismatic, with large basal nuclei which are generally spherical and poor in chromatin. The shapes of the cells and the positions of the nuclei obviously vary when the epithelium is pseudostratified. When the intermediate lobe is formed by cellular cords, the cells are more or less ovoid, with large, spherical, transparent nuclei; depending on the species, a variable number of small involuted cells lie at the centre of these cords. The cells of the intermediate lobe are always as large as or larger than any in the distal lobe, and this greater size is even more marked at the nuclear level. Apart from the special case of the burrowing reptiles, the only exception to this rule is *Crocodylus niloticus*, in which the intermediate lobe is large, but the cells appear poor in cytoplasm and show little chromophilia.

Table 2.3. Histochemical reaction and staining properties of the intermediate and gonadotrophic LH cells

	Intermediate cells					Gonadotrophic LH cells					
	PAS	AB-PF	E	A	C	PAS	AB-PF	H	E	A	C
Sphenodontidae	±	±	0	+	0	++	+	+	0	+	0
Gekkonidae	++	±	0	±	±	+++	±	+	0	+	0
Xantusiidae	++	+	0	±	±	+++	0	+	0	+	0
Delma fraseri	++	+	0	±	±	+++	±	+	0	+	0
Lialis burtonis	++	0	0	0	+	+++	0	0	+	0	0
Iguanidae	+	++	0	+	0	±	±	±	0	+	0
Anolis	0	+++	+	0	0	±	±	±	0	+	0
Agamidae	+	+	0	+	0	+	+	+	0	+	0
Chamaeleonidae	++	++	0	+	0	+	+	+	0	+	0
Lacertidae	+	±	0	±	±	++	±	+	0	+	0
Teiidae	±	±	0	±	±	++	0	+	0	+	0
Gerrhosaurinae	++	0	±	±	0	+++	0	±	0	+	0
Cordylinae	+	±	0	±	±	+++	0	±	0	+	0
Scincidae	+	+	0	0	+	+++	0	+	0	+	0
Feyliniidae						++	0	+	0	+	0
Anguidae	+	+	0	0	+	+++	+	+	0	+	0
Anniellidae						+++	±	±	0	+	0
Helodermatidae	±	±	0	0	+	+++	+	+	0	+	0
Varanidae	0	0	0	±	±	0	0	+	0	0	+
Trogonophidae	+	+	0	±	±	+++	+	0	0	0	+
Amphisbaenidae						+++	++	+	0	+	0
Typhlopidae						++	+	+	0	+	0
Leptotyphlopidae						++	++	+	0	0	+
Boidae	++	++	0	0	+	++	++	+	0	+	0
Colubridae	+	++	+	0	0	+	++	+	+	0	0
Elapidae	+	++	+	0	0	+	++	+	+	0	0
Hydrophiidae	+	++	+	0	0	+	++	+	+	0	0
Viperinae	+	++	0	0	+	+++	+++	+	0	+	0
Crotalinae	+	++	0	0	+	+++	+++	+	0	+	0
Testudines	±	0	0	±	0	+++	±	+	0	+	0
Crocodilia	+	+	+	0	0	+++	0	0	+	0	0

A. Amphophily: the secretory granules are both erythrophilic and cyanophilic; they stain violet with azan.

AB-PF. Cells stain, other permanganate oxidation, with alcian blue at pH3 and with paraldehyde fuchsin.

C. Cyanophily: an affinity for acid stains with low coefficients of diffusion, such as aniline blue and light green.

E. Erythrophily: an affinity for acid stains with high coefficients of diffusion, such as erythrosin, azocarmin, orange G, and azorubin S.

H. Cells stain with lakes of hematoxylin.

PAS. Cells give positive reaction with periodic acid-Schiff reagents.

+++, strong reaction; ++, moderate reaction; +, slight reaction; ±, variable reaction; 0, no reaction.

In the fossorial Boidae that have been studied and in *Trogonophis wiegmanni*, most of the cells of the intermediate lobe are involuted and lack secretory products, but a small number of them appear normal and well developed. In the Anniellidae all the intermediate cells are small, but still have chromophilic granules. In the other burrowing reptiles examined there are no functional cells in the intermediate lobe.

In most other cases, the entire supranuclear region of the cells of the pars intermedia is filled with fine granules; these are not very dense and are cyanophilic or somewhat amphophilic. They stain lightly with periodic acid Schiff (PAS) and retain a little paraldehyde fuchsin, alcian blue, or lakes of hematoxylin after permanganate oxidation. In *Anolis*, the Boidae, and the Colubroidea, the numerous secretory granules are large, very chromophilic, and stain strongly with paraldehyde fuchsin and alcian blue; in *Anolis*, the Colubridae, Elapidae, and Hydrophiidae, they are peculiar in being exceedingly erythrophilic.

Distal Lobe

The general form of the distal lobe is that of an anteroposteriorly elongated and more or less dorsoventrally flattened strip. As already noted, the pars distalis generally lies ventral to the pars nervosa and posteroventral to the median eminence; the hypophysis is bilaterally symmetrical. In the non-burrowing snakes, the distal lobe is situated ventrolateral to the neural lobe, which distinctly depresses its left

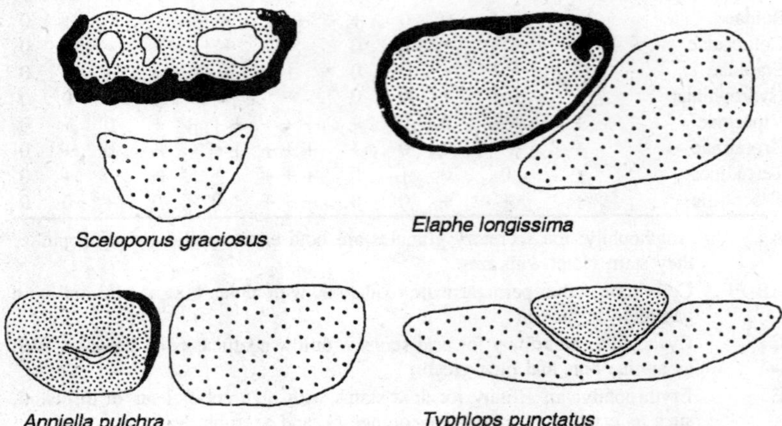

Fig. 2.2. Transverse sections through the hypophysis of Squamata. The neural lobe and regions with many neurosecretory fibres are indicated by fine stippling, the distal lobe by coarse stippling, and the intermediate lobe by black

dorsal region. In the Amphisbaenia and the Typhlopidae, the anterior part of the pars nervosa and the pituitary stalk occupy a longitudinal median depression in the pars distalis, the lateral regions of which form two flanges. In the Gekkota the morphology of the distal lobe is quite distinctive, with a well-developed anterior region and a flattened posterior region which becomes thickened again posterior to the intermediate lobe. In the Varanidae the distal lobe, which is lodged in the deep sella turcica, is massive and particularly thick. In the Boidae, Viperidae (including the Crotalinae), and *Acanthophis antarcticus*, the distal lobe is ovoid with a swollen posterior region. In *Anolis*, on the other hand, it is particularly elongate and small relative to the rest of the hypophysis. The distal lobe is very flat dorsoventrally in the Testudines, and has a regular ovoid form in the Crocodilia.

The pars distalis is composed of fairly regular and thick cellular cords, which are surrounded by a thin connective tissue membrane; these cords contain many chromophilic cells peripherally and many apparently involuted or undifferentiated chromophobic cells centrally. The proportion of these two cell types evidently varies during the functional cycle, but it is also an extremely constant specific characteristic. There are numerous capillaries, as well as vascular sinuses of varying size and form, between the cellular cords, but the capillaries are always much larger and more abundant in the posterior region. The vascular sinuses are almost absent in the Varanidae, but are consistently very large and numerous in the Anguidae, Helodermatidae, and Testudines.

The recognition and functional significance of the different types of cells within the pars distalis raise critical problems and thus require discussion. For a long time, aside from the chromophobic cells, only acidophilic cells and so-called "basophilic" cells were recognized. Then Siler (1936), Poris and Charipper (1938), Poris (1941), Hartmann (1944), Cieslak (1945), Miller (1948), and Wingstrand (1951) recognized two different types of erythrophilic cells in numerous squamates. Some, in the posterior region, generally contain fairly fine granules which colour orange with azan; others, in the anterior region, have large carminophilic granules. The former are one particular type of cell termed the alpha cell, but the latter, in most species, represent at least two types of cells.

At the present time the use of simple histochemical reactions and of some specific stains permits the recognition, in the pars distalis of reptiles, of five different and well characterized cellular types, or even six it the chromophobic or principal cells are included. In a few species (*Testudo graeca*, *Vipera aspis*, *Cerastes cerastes*, *Anguis fragilis*, and *Agama impalearis*), study of the changes in these different types of cells during the annual cycle and under certain experimental conditions provides quite precise information on the roles of three of the types. Rather strict and constant localization of each of these types permits their easy recognition in different species despite some variation in their histochemical characteristics and staining properties. However, attributing the same functional significance to them is clearly only a working hypothesis. In any case, the respective roles of the different types of adenohypophyseal cells need not be discussed in a morphological study. However, to avoid further complication of an already confused subject, I shall, wherever possible, use a functional nomenclature to define different types of cells in the pars distalis.

In the few cases which have been studied, the appearance of the different types of adenohypophyseal cells shows important variations during the course of the annual, and especially the reproductive, cycle mainly in females. This is clearly taken into account in the comparisons between species and families, comparisons which are based, as much as possible, on males.

The gonadotrophic LH (or gamma) cells are always localized in the anterior part of the pars distalis. Generally fairly numerous, these cells are prismatic or sometimes ovoid, and are often arranged in palisades along the capillaries; they have spherical basal nuclei. The secretory granules, of varying size and number, are amphophilic; that is, they stain simultaneously with aniline blue and the erythrosin dyes and thus appear violet in azan. The granules show a marked affinity for the lakes of hematoxylin and, to a varying degree, for alcian blue and paraldehyde fuchsin; they are generally strongly PAS-positive. In the Gekkonidae, the gonadotrophic LH cells are ovoid and scattered. In the Iguania and Lacertoidea they are small and rare; on the other hand, they are particularly large and numerous in the Anguidae, Helodermatidae, Colubridae, Elapidae, Hydrophiidae, and Crocodilia. In the latter four groups, they are clearly carminophilic, but they stain very lightly with PAS in the Colubroidea and their allies. Their constant location, amphophilia, and affinity to the lakes of hematoxylin permit the easy recognition of the gonadotrophic LH cells.

The gonadotrophic FSH (or beta) cells, less numerous than the preceding type, generally lie in the medial and lateral parts of the distal lobe, where they frequently form easily visible groups. However, they can also be dispersed throughout the anterior two-thirds of the pars distalis. They are generally ovoid cells, with small, spherical, central nuclei or, sometimes, with flat basal ones. The secretory product forms fine cyanophilic granules which stain to a variable degree with alcian blue and paraldehyde fuchsin and are PAS-positive. In some species, there are also large erythrophilic secretory granules or small erythrophilic lumps which stain strongly with alcian blue and paraldehyde fuchsin; in these cases the gonadotrophic FSH cells may be confused with thyrotrophic cells. In the Testudines, the numerous gonadotrophic FSH cells are distinctly fusiform. They are particularly large and easily recognized in some Elapidae and Viperidae.

The thyrotrophic (or delta) cells vary less among reptiles than do the other types of cells. They are rarely numerous, and may be scattered throughout the pars distalis or lie only in its posterior half. They are conical and often lie near the centres of the cellular cords, with long apical extensions towards the capillaries. They have spherical, basal nuclei. Their large or very large secretory granules may be either cyanophilic or erythrophilic; they always stain deeply with alcian blue and paraldehyde fuchsin, and are PAS-positive. In the Scincidae, Anguidae, and Helodermatidae, the thyrotropic cells are more or less cuboidal with irregular basal nuclei, and are arranged along the blood sinuses of the posterior part of the distal lobe. In the Colubridae, Elapidae, and Hydrophiidae, these cells are particularly large and numerous; they all appear to be erythrophilic, whereas in other forms having large granules of this type, the granules are separated from a cyanophilic substratum. In turtles, the thyrotrophic cells are larger than the gonadotrophic FSH cells.

The alpha cells, the function of which is unknown, are always numerous and restricted to the posterior half of the pars distalis. However, their form and size vary greatly from one family to another. They are generally prismatic or cuboidal with hemispherical or irregular basal nuclei, but may be lanceolate with ovoid central nuclei. The large, fairly regular secretory granules are coloured orange by azan, but are not stained by alcian blue or paraldehyde fuchsin; they are PAS-negative. The amount of granulation varies greatly both from one family to another and from one individual to another. When the alpha cells lack granules completely, they may appear mauve in azan.

The X cells, generally less numerous than the preceding ones, occur only in the medial or anteromedial parts of the pars distalis, where they often form small homogeneous cellular cords. In most reptiles they are large, prismatic or elongate ovoid cells, with ovoid basal nuclei. The fairly large, abundant secretory granules are distinctly carminophilic in azan and are not stained by alcian blue or paraldehyde fuchsin. In *Sphenodon punctatus*, the Chamaeleonidae, and the Colubroidea, these granules react slightly with PAS. In the Anguidae, Helodermatidae, and Ophidia, the fairly small X cells are greatly reduced in number; moreover, their staining properties are very similar to those of the alpha cells. The X cells often tend to be clustered around small pseudofollicles composed of lanceolate cells with elongated central nuclei. Granules are restricted to the supranuclear region.

The so-called chromophobic or principal cells, small cells lying in the centre of the cellular cords, do not represent a separate category. In most cases, close examination reveals some secretory granules which relate the cells to one or another of the different chromophilic types; the chromophobic types simply represent a stage in their involution. Others are apparently undifferentiated, but the massive hyperplasia of many types of cells, particularly of the gonadotrophic cells at certain stages of their cycle, suggests that the chromophobic cells may become functional. Whether they can be transformed into any cellular type is still being discussed, but the strict localization of the different types of cells in the pars distalis in reptiles does not support this hypothesis. The abundance of involuted or apparently undifferentiated cells varies with the stage of the secretory cycle, and is also a very constant specific character. These cells are particularly rare in the Anguidae, Helodermatidae, and, to a lesser degree, Scincidae. On the other hand, they are very numerous in the Iguania and Lacertoidea.

Histochemical studies in a strict sense have only rarely been carried out on reptilian adenohypophyses. The intracellular secretory granules are always rich in proteins, but the amount of sulphydrylated protein varies with the species and type of cell. A positive reaction with PAS indicates the presence of mucoproteins and is characteristic of the so-called "mucoid" cells, that is the gonadotrophic and thyrotrophic elements; acid mucopolysaccharides are apparently lacking. The few data on reptiles do not permit a useful discussion of the histochemical characteristics of the different types of adenohypophyseal cells. This problem has been treated, for vertebrates in general, in several reviews, notably that of Herlant (1962) which contains an extensive bibliography.

The size of the different cells in the pars distalis evidently varies according to their type and to their stage in the functional cycle, but, taking these factors into account, it is also a specific character. They are particularly large, with surface areas of about 100 μ^2, in *Sphenodon punctatus*, the Anguidae, the Helodermatidae, *Tiliqua scincoides*, and, to a lesser degree, the Scincidae, Boidae, and Colubroidea. On the other hand, they are relatively small in the Iguania and Lacertoidea, where their average surface area is less than 50 μ^2. In the burrowing reptiles, the adenohypophyseal cells are much smaller than those in related families, but the relative differences between the different systematic groups persist. The cells of the Leptotyphlopidae are particularly small and have a very sparse cytoplasm.

The form of the cells of the pars distalis depends more on their type than on the systematic position of the animal. However, in the Testudines all the cells tend to be characteristically elongated and fusiform.

The proportion of the different types of chromophilic cells in the distal lobe, without considering some undifferentiated cells, varies with the species studied and the stage of the functional cycle. There are no systematic numerations, but the following very approximate proportions can be recognized: alpha cells, 25 to 40% (perhaps 50% in the Agamidae); gonadotrophic LH cells, 20 to 35%; gonadotrophic FSH cells and thyrotrophic cells 10 to 20% each; X cells, 5 to 25%.

The relative volume occupied by each cellular type, irrespective of their numbers and aside from seasonal variation, may differ greatly from one family to another. In the Colubridae, Elapidae, Hydrophiidae, Anguidae, and Helodermatidae, the gonadotrophic LH cells fill most of the anterior two-thirds of the pars distalis, and can occupy almost half its total volume. In contrast, the alpha cells dominate in the Agamidae and Varanidae.

In some mammals, a sixth type of cell, which appears to secrete corticotrophic hormone (ACTH), can be demonstrated after preliminary treatment of the animal. They probably also occur in reptiles in which such cells could easily be confused with more or less degranulated alpha cells.

Pars Tuberalis

In the Rhynchocephalia and Crocodilia, the pars tuberalis consists of small groups of cells forming pseudo-follicles along the ventral surface of the infundibular recess, especially in the vicinity of the median eminence. These cells, which are slightly larger in the Crocodilia than in *Sphenodon punctatus*, have spherical central nuclei and chromophobic or slightly erythrophilic cytoplasm.

In the Testudines the pars tuberalis is continuous with the anterodorsal part of the pars distalis. Besides the numerous small chromophobic cells, there are some large ovoid cells filled with large granules which are moderately PAS-positive and stain strongly with alcian blue and paraldehyde fuchsin. These cells, which resemble slightly the thyrotrophic cells, are also found in very small numbers in the pars tuberalis of *Sphenodon punctatus*.

The Squamata generally lack a pars tuberalis, but sometimes traces of it may be found beside the median eminence in the form of a thin layer of visibly involuted, small chromophobic cells.

Embryonic Development and Vascularization

Morphogenesis

Numerous works describe the development of the reptilian hypophysis, which is essentially similar to that in other amniotes; it involves the simultaneous participation of an oral ectodermal bud and of a primordium from the diencephalon.

In the Sauria, a depression appears in the roof of the mouth at the 22-27 somite stage. This primordium of Rathke's pouch deepens rapidly, and its blind, aboral extremity comes into contact with the floor of the diencephalon. A little later, a cavity extends posteriorly within the infundibulum at the level of contact with Rathke's pouch, gradually bending the latter posteriorly. The walls of Rathke's pouch thicken, and the pouch develops a stalk by the contraction from its oral connection. The aboral extremity of the pouch becomes slightly more posterior, and widens out at its contact with the diencephalon. At the same time a diverticulum is formed anteriorly, and two cellular cords, the Anlagen of the pars tuberalis, extend laterally. These three structures extend progressively towards the diencephalon and become connected to it, the first at the level of the future median eminence and the other two slightly more dorsolaterally.

Next, the hypophyseal canal closes completely, and the lumen of Rathke's pouch is gradually filled by the formation of cellular cords. Thus the adenohypophysis consists of an anterior (oral) lobe formed from two narrow, elongate lateral expansions and of a posterior (aboral) lobe, which is enlarged at its posterior end. This enlarged area becomes the intermediate lobe, which is more or less distinct depending on the species; the anterior lobe and the anterior part of the posterior lobe fuse to form the pars distalis. However, as Wingstrand (1951) has demonstrated in both birds and reptiles, the adenohypophyseal cytology, even in the adult, shows traces of this initial separation. The infundibular recess, originally elongated posteriorly, swells at its posterior end, but the pituitary stalk remains largely open. It is at this stage that the hypophyseal portal system, which will provide most or all of the blood supply to the distal lobe of the adult, is established.

During the second half of embryonic development, the hypophysis slowly assumes its definitive form. The neural lobe enlarges and differentiates, the intermediate lobe becomes even further separated from the anterior part of the original posterior lobe, and the pars distalis completes its differentiation. The hypophyseal cleft is reduced and may disappear completely within the distal lobe, but it persists within the intermediate lobe where it extends around the neural lobe to a variable degree. Finally, the pars tuberalis atrophies and loses its contact with the distal lobe; at birth it is represented only by two cellular islets attached to the floor of the diencephalon on either side of the median eminence.

The hypophyseal development in other reptiles hardly differs from that which has just been described. However, in the Ophidia, the distal lobe takes up, well after birth, a position lateral to the mass formed by the complex of the neural and intermediate lobes. The atrophy of the pars tuberalis seems more marked than in lizards; on the other hand, the pars nervosa continues to thicken during the last weeks of embryonic development and only in the adult assumes the large and full form already described. In *Sphenodon*, the Crocodilia, and especially the Testudines, there is no involution of the pars tuberalis; even when the latter loses its lateral contact with the distal lobe, well-developed cellular cords arc retained on the infundibular floor.

All recent workers agree that the diencephalic floor plays an inductive role in the differentiation of the intermediate lobe. In *Chamaeleo bitaeniatus*, a species with a marked ability of changing colour and a hypertrophied intermediate lobe, the area of attachment of the base of Rathke's pouch to the infundibular floor is more extensive in the earliest stages than in the Scincidae and Lacertidae; this characteristic persists throughout life.

At birth, the morphology of the hypophysis is quite similar to that of the adult, except that the posterior part of the pars distalis is proportionally smaller. In *Vipera aspis*, the only species in which this phenomenon has been studied in detail, the development of this region occurs during the first weeks of life, rather than at the time of sexual maturity.

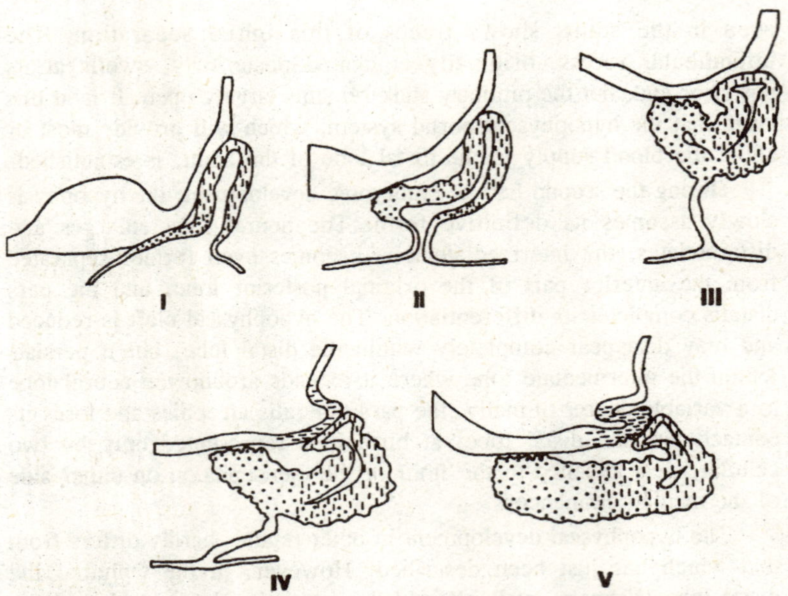

Fig. 2.3. Embryonic stages in the development of the hypophysis of Natrix natrix.

Cytogenesis

Although the morphology of the hypophysis of numerous reptiles of all orders and most suborders has long been studied, the differentiation of the various glandular cells during embryonic growth is still very poorly known, having been described in only two works, one on a snake and one on a lizard.

In *Vipera aspis*, secretory granules first appear in the adenohypophysis at about the middle of gestation, shortly after the establishment of a vascular connection between the median eminence and the anterior end of the pars distalis. Both thyrotrophic cells posteriorly and gonadotrophic FSH cells medially and laterally can be recognized in the well formed distal lobe, which already has a hypophyseal cleft. A little later, the gonadotrophic LH cells differentiate in the anterior part of the pars distalis. In the second third of gestation some erythrophilic cells, the X cells, are recognizable in the anteromedial region; 8 to 15 days later secretory products appear in the cells of the intermediate lobe. Only at the end of embryonic development, or sometimes even after birth, do cells of the last adenohypophyseal type, the alpha cells, differentiate in the posterior region which is then developing rapidly. The secretory granules of the adenohypophyseal cells have, from their first appearance, the same staining properties and histochemical characteristics as in the adult.

At birth, most of the differentiated adenohypophyseal cells show signs of functional activity and are more or less non-granular and vacuolated. This pattern is retained until the onset of hibernation; then many cells, which had been chromophobic, differentiate. The numbers of alpha cells and, to a lesser degree, thyrotrophic cells increase the most. For two to three years, until the approach of sexual maturity, the histology of the distal lobe changes very little; the gonadotrophic FSH cells are quiescent, the gonadotrophic LH cells are clearly involuted, the numerous large X cells are apparently active, and, as in the adults, the alpha cells are first large and filled with secretory granules and later become more or less non-granular and of variable size.

In *Xantusia vigilis*, which has an embryonic period of about 13 weeks, the first secretory granules, partially PAS-positive, appear in the adenohypophysis of 8 mm (about 5 week-old) embryos. By the middle of the gestation period, in 9 to 10 mm embryos, the gonadotrophic LH (gamma) cells anteriorly, the gonadotrophic FSH (beta) and thyrotrophic (delta) cells posteriorly and medially, and the cells of the intermediate lobe are clearly differentiated. However, although PAS-positive products are fairly abundant, no secretory granules stain with paraldehyde fuchsin. Later the different types of cells come to resemble those of the adult. The alpha cells differentiate only about two weeks before birth; as in some Gekkonidae, X cells and alpha cells cannot be distinguished even in adults.

The results obtained in this lizard and the snake *Vipera* are therefore very similar; however, definitive secretory granules tend to appear at slightly earlier stages in *Xantusia vigilis*.

The cytogenesis of the hypothalamo-neurohypophyseal tract is even less well known than that of the adenohypophysis. In *Vipera aspis*, neurosecretory products detectable by paraldehyde fuchsin, alcian blue, or chromic hematoxylin, after permanganic oxidation first appear in very small quantities in the median eminence, and their appearance is noted in the neural lobe at the end of the second third of gestation, at about the same time as secretion appears in the gonadotrophic LH cells and the X cells. Secretory products only become abundant in the pars nervosa during the first hibernation.

In *Xantusia vigilis*, neurosecretory products appear in the median eminence and in the neural lobe in 9 to 10 mm embryos (near the middle of the gestation period), and their quantity increases noticeably until birth. *Xantusia* and *Vipera aspis* thus differ considerably, but it should be remembered that neurosecretory products are generally much more abundant in lizards than in snakes.

Vascularization

The physiological importance of the blood supply to the hypophysis has provoked numerous anatomical studies; those of Green (1951), Wingstrand (1951), and Diepen (1952) deal with reptiles. More recently, Enemar (1960) has studied the development of the vascular system in some squamates. This last work contains an extensive bibliography.

In the Reptilia, the blood supply to the hypophysis is fundamentally the same as in the other amniotes. Its essential characteristic is the presence of a portal system which supplies the distal lobe with venous blood from the capillaries of the primary plexus in the median eminence.

In lizards (*Anolis carolinensis*, *Anguis fragilis*) the hypophysis appears to receive all its blood from the infundibular arteries, which arise from the anterior ramus of the internal carotid. The infundibular artery divides into an anterior branch, which extends to the vicinity of the optic chiasma, and a posterior branch, which is nearer the hypophysis and provides its major blood supply. At the level of the median eminence, the capillaries covering the infundibular floor form a very dense primary plexus which does not penetrate deeply into the neural tissue. The capillaries unite to form four to six rather elongate portal vessels, thus providing the connection with the anterodorsal part of the pars distalis. There they form a secondary plexus which supplies the entire distal lobe. Capillaries from the primary network of the median eminence vascularize the pars nervosa and the pars intermedia; these capillaries do not penetrate the neural lobe, but, in *Anolis carolinensis*, lie between the cellular cords of the hypertrophied intermediate lobe. The retrohypophyseal vein receives all the blood from the hypophysis.

In snakes (*Thamnophis sirtalis*, *Natrix natrix*, *Vipera berus*), the primary plexus of the medium eminence is particularly well developed, and the portal vessels are often more numerous (6-8) and larger than in lizards. The neural lobe receives a few capillaries from the median eminence, but most of its blood supply is from arterioles originating directly from the infundibular arteries. The capillaries do not penetrate the neural tissue, but accompany the connective tissue septa which divide it into many lobules; they also surround the intermediate lobe. Because of the large area of contact between them, the pars intermedia undoubtedly receives, especially posteroventrally, some blood from the

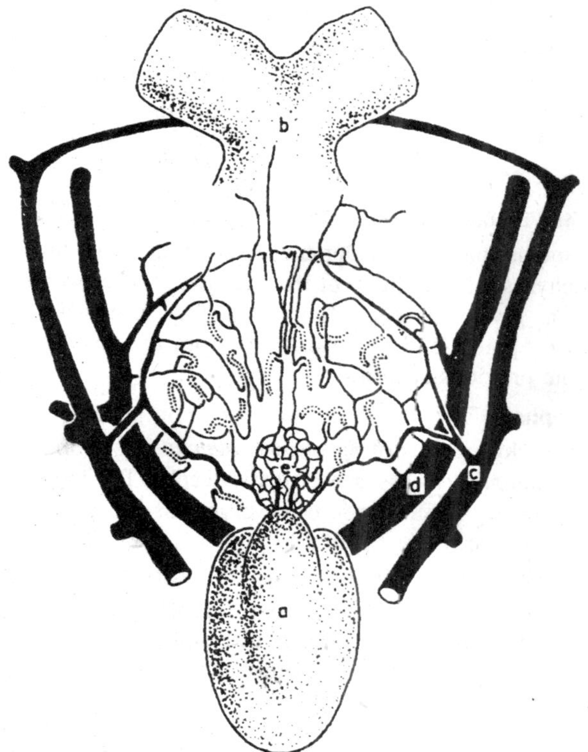

Fig. 2.4. Lacerta agilis (Lacertidae). Ventral view of the blood vessels of the hypothalamus between the optic chiasma and the hypophysis. a, Hypophysis; b, optic chiasma; c, anterior ramus of the internal carotid branching off the infundibular artery; d, vein draining the hypophysis; e, primary plexus on the median eminence.

pars distalis. Most of the blood supply to the distal lobe is provided by the important hypophyseal portal system which joins the anterior end of the distal lobe to the median eminence; however, the pars distalis also receives some arterioles directly from the internal carotids.

The vascularization of the hypophysis of turtles (*Chrysemys picta*, *Testudo graeca*) and crocodilians (*Alligator mississippiensis*) resembles closely that of snakes. However, in turtles and crocodilians, the portal must cross the cellular cords of the pars tuberalis before reaching the anterodorsal surface of the pars distalis.

Inter-carotid anastomoses, which are characteristic of birds, occur in the Testudines and the Crocodilia. They are absent in the Lepidosauria and mammals. Posterior to the distal lobes, the cavernous sinuses of turtles and crocodilians are replaced by a simple enlarged transverse vein in the Lepidosauria.

The positions and courses of the larger vessels have been little studied in reptiles, but the essential point—that the pars distalis is supplied with venous blood originating from the median eminence—has been well demonstrated in all the orders.

Comparative Morphology of the Reptilian Hypophysis

It is impossible to consider the detailed anatomy and cytology of the hypophyses of all the reptiles that have been studied. A brief summary is given for each family, paying particular attention to characteristic points and to differences from the general descriptions given in the previous sections.

Rhynchocephalia

The neurohypophysis of *Sphenodon punctatus* is of the saurian type, and the infundibular recess penetrates it deeply. The median eminence is characterized by the neurosecretory fibres passing through the ventral limiting glia towards the capillaries.

The intermediate lobe surrounds the posterior two-thirds of the distal lobe. It is made up of the two typical layers; as in all large reptiles, the internal layer is festooned and largely formed by true cellular cords. The tall, prismatic intermediate cells contain fine, barely chromophilic granules.

The massive distal lobe is connected to the intermediate lobe only by a very narrow bridge. The gonadotrophic LH cells show no special peculiarities. The gonadotrophic FSH cells, scattered throughout the distal lobe, are particularly rare and weakly chromophilic. The

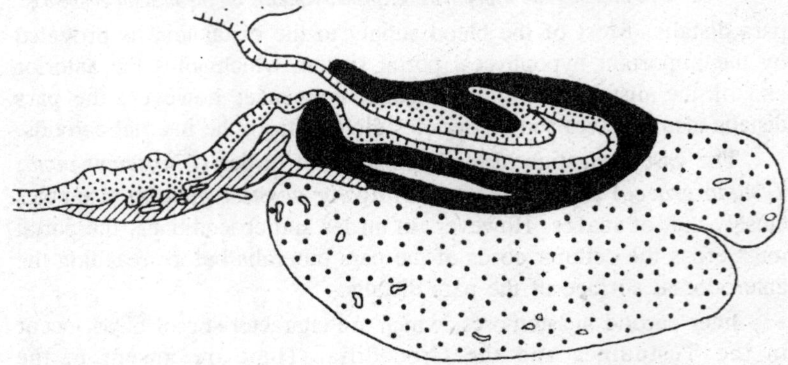

Fig. 2.5. Sagittal section through the hypophysis of Sphenodon punctatus.

thyrotrophic cells, which are more numerous than the former, are also scattered throughout or sometimes clustered in small groups; they contain large granules or a few small erythrophilic particles on a cyanophilic base. The very numerous alpha cells lack granules in the posterior third of the pars distalis, but contain many fairly large ones in the central third. The X cells show the same staining properties as the alpha cells, but are easily distinguished by their elongate ovoid form, their location in small homogeneous cellular cords at the anterior end of the pars distalis, and their positive reaction to PAS.

Sauria

Gekkonidae

The median eminence and the neural lobe show no peculiarities. The large intermediate lobe almost completely surrounds the pars nervosa, and its two layers are separated by a continuous hypophyseal cleft. The epithelium of the internal layer is simple and formed by high columnar cells; the external layer is squamous.

The distal lobe, which is attached to the posteroventral side of the intermediate lobe, has a characteristic form with a well-developed anterior region and a flattened posterior region. The relatively few gonadotrophic LH cells are ovoid and scattered throughout the anterior half. The gonadotrophic FSH cells, often prismatic, are more numerous in the central region and contain fine cyanophilic granules. The few thyrotrophic cells are scattered throughout the lobe and are prismatic or conical with numerous large granules. The alpha cells, localized in the flat posterior region, are particularly small and few in number; they are generally only slightly granular. The larger X cells, which are fairly numerous in the anterior region, contain numerous, medium-sized secretory granules.

Pygopodidae

The hypophysis of *Delma fraseri* is entirely like those of the Gekkonidae, but that of *Lialis burtonis* shows a number of peculiarities. The sella turcica of the latter is shallower, and the complex of neural and intermediate lobes tends to lie posterior to the distal lobe and on almost the same horizontal plane, although the posterior part of the pars nervosa is completely flat. Moreover, the median eminence has a narrow attachment to the diencephalon. The intermediate lobe is well developed, and its festooned internal layer tends to form cellular cords. Except in this last character, the hypophysis of *Lialis burtonis* suggests the condition in fossorial lizards. The gonadotrophic LH cells are

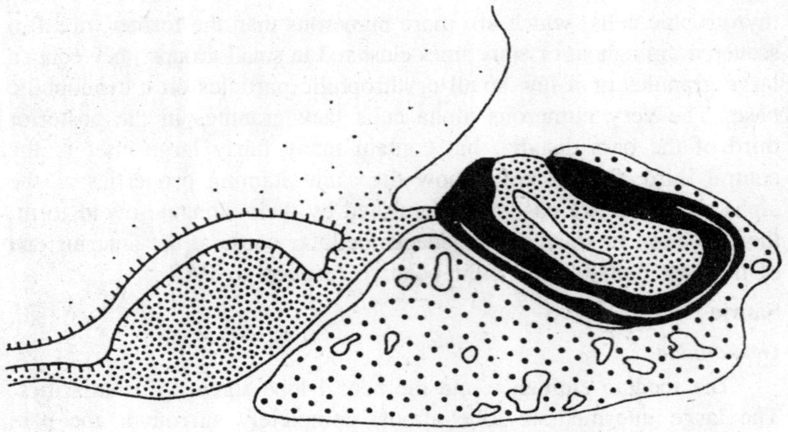

Fig. 2.6. Sagittal section through the hypophysis of Delma fraseri.

large and numerous, do not retain hematoxylin lakes, and are strongly erythrophilic. However, the hypophysis of *Lialis* basically resembles that of the gekkonids.

Xantusiidae

The hypophyses of *Xantusia vigilis* and *X. henshawi* closely resemble those of the Gekkonidae and *Delma fraseri*. However, the intermediate lobe is a little less developed and slightly less flat posteriorly, and the alpha cells, which are larger and contain many secretory granules, hardly differ from the X cells.

Iguanidae

The neural lobe is fairly large, irregularly shaped, and deeply penetrated by the infundibular recess. The small intermediate lobe caps the posterior third or half of the pars nervosa; it is formed by small, irregular cellular cords containing numerous involuted cells.

The form of the rather small distal lobe varies. The often parallel cellular cords are separated by numerous, elongate, very small vascular sinuses and contain many undifferentiated cells. The few gonadotrophic LH cells are restricted to the anterior end; they are small and weakly chromophilic. The more numerous gonadotrophic FSH cells are concentrated ventromedially; they contain fine secretory granules and sometimes some small, regular particles. The fairly numerous thyrotrophic cells are scattered throughout the gland and always have large dense granules. The moderate-sized alpha cells contain fairly fine and rather weakly orangophilic granules, and are abundant posteriorly. The X cells, which are a little larger than the preceding ones and may also be numerous, are restricted to the anterior half of the lobe; their secretory granules are fairly large, strongly staining, and carminophilic. Generally the adenohypophyseal cells are small; the nuclei, which are often irregular but poor in chromatin, are also small.

The hypophysis of *Anolis* is characterized by a considerable hypertrophy of the intermediate lobe. The latter, which represents more than two-thirds of the hypophyseal volume, is made up of thick cords formed of large, regularly arranged prismatic cells; their very abundant secretory product appears as large erythrophilic granules which are PAS-negative, but stain strongly with alcian blue and paraldehyde fuchsin. The distal lobe is reduced to a long, thin strip lying ventral to the intermediate lobe.

Agamidae

The neural lobe is quite small and narrow. The well-developed intermediate lobe, formed from numerous, fairly thick cellular cords, almost completely encloses it. The tall, prismatic intermediate cells contain fine, weakly chromophilic granules.

The rather elongate and narrow distal lobe is slightly enlarged anteriorly and is formed from thick, irregular cellular cords with many undifferentiated cells. The rare gonadotrophic LH cells are restricted to the anterior end and contain many, rather large, weakly chromophilic granules. The gonadotrophic FSH cells are common in the anterior and anteromedial regions and have fine, weakly chromophilic granules. The conical thyrotrophic cells are scattered throughout the gland and contain fairly large cyanophilic granules. The alpha cells are generally non-granular and small; they occupy most of the posterior two-thirds of the pars distalis. The X cells are larger and much less numerous than the preceding ones; they have fairly large carminophilic granules and form rare, almost homogeneous cellular cords in the anterior region. The nuclei of the adenohypophyseal cells are generally small and irregular.

Chamaeleonidae

The morphology of the hypophysis is extremely homogeneous in the known representatives of this family; there is a moderate-sized, slightly flattened neural lobe, a well-developed intermediate lobe which extends mainly lateral and posterior to the neural lobe, and a massive distal lobe. The cords of the intermediate lobe are regular, with large prismatic cells containing fine weakly chromophilic granules.

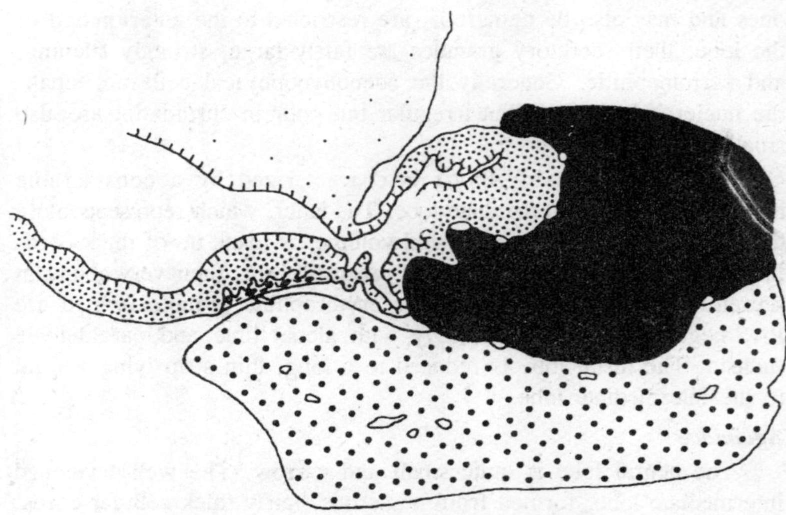

Fig. 2.7. Sagittal section through the hypophysis of Chamaeleo lateralis.

As in the two preceding families, the few gonadotrophic LH cells are weakly chromophilic and restricted to the anterior end of the pars distalis. The numerous gonadotrophic FSH cells are scattered throughout the anterior two-thirds, but are distinctly commoner ventrally; they contain fine granules. The few scattered thyrotrophic cells are small and have large cyanophilic granules. The rather large alpha cells are common posteriorly and contain a variable number of large orangophilic secretory granules. The X cells, concentrated in the medial and anteromedial regions, have large carminophilic granules which are peculiar in being strongly PAS-positive. The undifferentiated cells at the centre of the cellular cords, while numerous, are slightly less common than in the two preceding families. The nuclei of the adenohypophyseal cells are large and poor in chromatin, but are very irregular and often divided.

Lacertidae

The small neural lobe is ovoid with an extended longitudinal axis. The poorly developed intermediate lobe surrounds the posteroventral half of the pars nervosa. It is made up of two or three rows of cells, variable in form, some being involuted while others contain rather weakly staining granules of moderate size. The hypophyseal cleft is often reduced.

The massive distal lobe is formed by thick cellular cords with many undifferentiated cells. There are hardly any blood sinuses. The few, highly chromophilic gonadotrophic LH cells are restricted to the anterior end of the lobe. The equally rare gonadotrophic FSH cells have rather weakly staining, very fine granules and generally occur in the ventromedial region. The thyrotrophic cells are smaller than the preceding ones and are scattered throughout the distal lobe; they contain a variable number of large cyanophilic granules. The alpha cells, which are very numerous posteriorly, are small and usually non-granular. On the other hand, the X cells, which are collected in homogeneous cords or sometimes in small, regular pseudo-follicles in the anteromedial region, are always filled with carminophilic secretory granules. The nuclei of the adenohypophyseal cells are generally small and irregular; although not really abundant, the chromatin is less scanty in lacertids, and indeed in most lacertoids, than in most other reptiles.

In *Acanthodactylus scutellatus* (but not in some other members of this genus) the distinctly hypertrophied intermediate lobe is formed by large, irregular cellular cords like those of other lacertids, but made up of much larger cells with more secretory products.

Teiidae

The neural and intermediate lobes are similar to those of the Lacertidae. However, the hypophyseal cleft is more distinct and, in the posteroventral part of the intermediate lobe, delimits the two characteristic layers.

The distal lobe is long and narrow. The irregular cellular cords contain many undifferentiated cells. There are fairly numerous, elongate vascular sinuses, particularly in the posterior region. The different types of cells of the pars distalis do not differ significantly from those of the Lacertidae. The gonadotrophic FSH cells resemble closely the thyrotrophic cells.

Cordylidae (including the Gerrhosaurinae)

The hypophyses of the few studied representatives of this family generally resemble those of the Lacertidae and the Teiidae. They differ in the greater number of gonadotrophic LH and FSH cells. The latter contain large granules similar to those of the thyrotrophic cells.

Scincidae

The hypophyses of the Scincidae differ markedly from all those that have already been described. The pyriform neural lobe is quite deeply penetrated by the infundibular recess. The intermediate lobe, which covers the posteroventral half and the lateral parts of the neural lobe, is made up of a high columnar internal layer of cells and a squamous external layer, except posteromedially where it is invaded by numerous alpha cells.

The elongate pars distalis has a fairly constant thickness and is formed by distinct cellular cords, which are rather thin and contain few undifferentiated cells. There are some small, elongate vascular sinuses. Gonadotrophic LH cells are numerous in the anterior half; they are large with moderate-sized granules and are arranged regularly along the cords. The ovoid or spherical gonadotrophic FSH cells are scattered throughout the anterior two-thirds of the lobe and contain both large cyanophilic granules and small erythrophilic particles. The cuboidal thyrotrophic cells, which are restricted to the posterior third or half of the pars distalis, are smaller than the preceding ones, but have an apparently identical secretory product. The orangophilic alpha cells are numerous in the posterior half; they are generally small and non-granular. The X cells, which form small homogeneous cords anteriorly, are always larger and densely filled with moderate-sized, carminophilic granules. In all species, the nuclei are rather small and often irregular and very poor in chromatin. In *Tiliqua scincoides* the adenohypophyseal cells are particularly large.

Feyliniidae

The wide neural lobe lies posterior to the distal lobe. The intermediate lobe is reduced to a thin band joining the posteromedial end of the neural lobe to the posterior end of the pars distalis; it has layers of small, apparently non-functional, cuboidal chromophobic cells. The distal lobe is enclosed between the median eminence and the pars nervosa. As in all limbless burrowing reptiles, the sella turcica is poorly developed and the hypophysis is extremely flattened dorsoventrally.

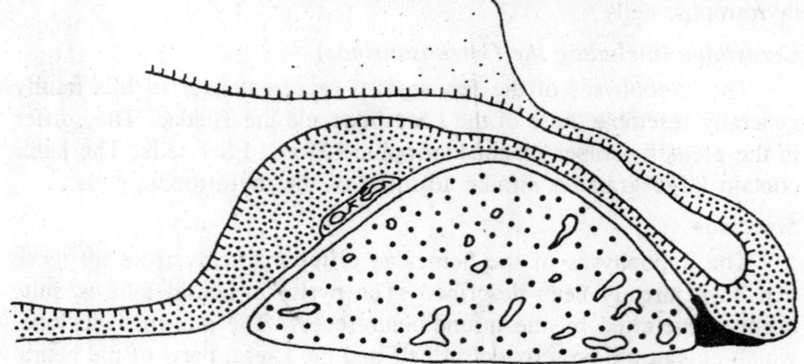

Fig. 2.8. Sagittal section through the hypophysis of Feylinia currori.

The cellular cords of the pars distalis are narrow and contain few undifferentiated cells. The different types of cells are poorly localized and difficult to recognize; the gonadotrophic FSH and thyrotrophic cells, and the alpha and X cells cannot be distinguished. The gonadotrophic LH cells are concentrated anterodorsally.

Anguidae

The neural and intermediate lobes resemble those of the Scincidae. The part of the large distal lobe posterior to the intermediate lobe is slightly incurved dorsally. Its cellular cords are distinct and almost completely devoid of undifferentiated cells. The numerous, very large vascular sinuses give a characteristic appearance to the posterior part of this lobe.

The many, very large gonadotrophic LH cells occupy much of the anterior half of the pars distalis and contain fairly fine but very densely staining granules. The gonadotrophic FSH cells, which are scattered throughout the anterior half, resemble those of the Scincidae. The same is true of the cuboidal thyrotrophic cells which are regularly arranged along the large blood sinuses of the posterior region. On the other hand, the more or less carminophilic alpha cells are large, prismatic, and almost always filled with moderate-sized granules. The rather few X cells are concentrated in small groups anteromedially; their granules resemble those of the alpha cells, but are distinctly less abundant. Most adenohypophyseal cells of the Anguidae are particularly large, as are the nuclei which are poor in chromatin and have very distinct nucleoli.

Anniellidae

The hypophysis of *Anniella* clearly shows adaptations to a fossorial life, but a little less than that of some other reptiles. The spherical neural lobe lies lateral to the distal lobe; the infundibular recess penetrates it deeply, but it is very flat and its lumen is almost completely closed. The poorly developed intermediate lobe is of the usual lacertilian form, and the cells of its internal layer appear functional. The ovoid distal lobe is formed from distinct cellular cords which are almost devoid of undifferentiated cells. There are numerous, very large blood sinuses in the posterior part of this lobe.

As in the Feyliniidae, only three types of chromophilic cells can be distinguished in the pars distalis, and they are not well localized. The nuclei are often more irregular than in the Anguidae, but the presence of large vascular sinuses in the distal lobe, the relatively large size of the cells, the scarcity of undifferentiated elements, and the histochemical characteristics of the gonadotrophic LH cells demonstrate the close relationship between the Anguidae and the Anniellidae.

Helodermatidae

The hypophyses of *Heloderma suspectum* and *H. horridum* are almost identical to those of the Anguidae, the only difference being that the intermediate lobe is a little smaller in *Heloderma*.

Varanidae

In contrast, the hypophyses of the Varanidae are completely different from those of the preceding families. The quite large neural lobe has many neurosecretory fibres and is largely surrounded by the voluminous intermediate lobe. The latter is made of thick cellular cords separated by connective tissue septa. The cords contain both weakly chromophilic polyhedric cells and distinctly involuted cells.

The thick, massive distal lobe is lodged in a very deep sella turcica. It is almost completely devoid of vascular sinuses. The gonadotrophic LH cells, which are restricted to the anterior third, are weakly chromophilic and PAS-negative. One type of cell is difficult to identify: rare, very chromophilic ovoid cells restricted to the ventromedial region. Other, slightly smaller cells are fairly numerous and scattered throughout the pars distalis; they contain few granules. The very numerous alpha cells are filled with orangophilic secretory granules and occupy much of the posterior and medial regions. The

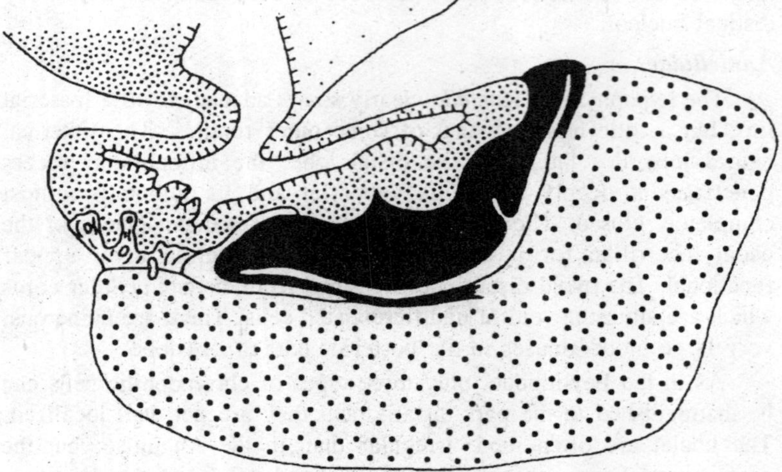

Fig. 2.9. Sagittal section through the hypophysis of Varanus niloticus.

abundant X cells, which are restricted to the anterior third, are often very elongate and contain dense carminophilic secretory granules. Most of the nuclei are fairly small, irregular, and poor in chromatin.

Amphisbaenia

Amphisbaenidae

The ovoid, elongate neural lobe lies posterodorsal to the distal lobe into which it is strongly pressed along a median groove; the extremely flat infundibular recess does not penetrate it very deeply. The well-developed median eminence lies near the distal lobe. The intermediate lobe is reduced to a long thin band ventral to the pars nervosa and appears to lack functional cells.

The large, columnar gonadotrophic LH cells are common in the anterior half of the pars distalis. The gonadotrophic FSH cells are scattered throughout the lobe and contain rather lightly staining, moderate-sized granules. The conical thyrotrophic cells are also scattered throughout the lobe and have more abundant secretory granules. The few small or very small alpha cells are found only at the posterior end of the lobe. The fairly large X cells are often elongate and filled with carminophilic granules; they occur medially and anteriorly. The nuclei generally contain more chromatin than those of most other reptiles.

Trogonophidae

The hypophysis of *Trogonophis wiegmanni* resembles that of the Amphisbaenidae, but its neural lobe is larger and more posteriorly situated in relation to the distal lobe which the pituitary stalk hardly depresses. The intermediate lobe is slightly less atrophied than in *Blanus cinereus* and contains some cells having secretory granules.

The quite rare gonadotrophic LH cells are restricted to the anterior third of the pars distalis; their many, moderate-sized secretory granules do not stain with hematoxylin lakes. The ovoid gonadotrophic FSH cells contain fairly fine, dense granules and are scattered throughout the lobe but are most numerous dorsomedially. The few thyrotrophic cells are scattered in the posterior and ventromedial regions; they contain large, dense granules. The alpha and X cells resemble those of *Blanus cinereus*. They also have similar nuclei.

Ophidia

Typhlopidae

The large, elongate neural lobe lies posterior to the distal lobe. The infundibular recess hardly penetrates it. The intermediate lobe seems to have disappeared completely. As in the Amphisbaenidae, the flat distal lobe is markedly depressed in the midline by the pituitary stalk and the entire anterior part of the neural lobe. The irregular and indistinct cellular cords contain only a few undifferentiated cells. There are no vascular sinuses.

The rather uncommon gonadotrophic LH cells are restricted to the anterior end of the pars distalis. The gonadotrophic FSH cells, which are particularly abundant medially and ventromedially, generally contain numerous secretory granules, but may sometimes be hypertrophied, more or less non-granular, and vacuolate. The numerous thyrotrophic cells are scattered throughout the gland and have large erythrophilic granules. The very small alpha cells occur only at the posterior end of the lobe. The X cells are distinctly larger than the alpha cells and are scattered in small groups throughout the distal lobe; they contain many large orangophilic granules. The nuclei of the adenohypophyseal cells of the Typhlopidae are a little larger and distinctly poorer in chromatin than those of the Amphisbaenidae.

Leptotyphlopidae

Aside from the flatness of the sella turcica and the absence of an intermediate lobe, the hypophysis of the *Leptotyphlops dulcis* shows no resemblance to that of the Typhlopidae. The neural lobe is reduced to a small spherical mass lateral to the distal lobe. The well developed or even hypertrophied median eminence is broadly attached to the enlarged anterior end of the pars distalis.

The distal lobe forms a thick band, truncated anteriorly and flattened posteriorly. It is characterized cytologically by the small size of the cells, all of which have very few secretory granules and are nearly filled by the large, turgid, transparent nuclei. The few gonadotrophic LH cells lie at the periphery of the cellular cords in the anterior region. Most of the gonadotrophic FSH cells, which are slightly larger than the preceding ones, are medially located. The thyrotrophic cells are uncommon, very elongate, and scattered throughout the lobe. The alpha, cells contain few large orangophilic granules and are dispersed throughout the distal lobe. The carminophilic X cells are restricted to the medial region.

Boidae

The massive, spherical neural lobe lies dorsolateral to the distal lobe in which it is partially enclosed. The infundibular recess does not penetrate it, but connective tissue septa incompletely divide it into lobules. The median eminence lies just dorsal to the anterior end of the pars distalis, and the pars tuberalis is less developed than in the Colubroidea. The quite irregularly shaped intermediate lobe encloses the posterior part of the neural lobe and rejoins the posterodorsal region of the right extremity of the pars distalis. It is made up of irregular cellular cords, in which the larger cells are filled with large very chromophilic granules.

The massive ovoid distal lobe has irregular thick cords containing some undifferentiated cells. Small vascular sinuses occur posteriorly. The numerous, very chromophilic gonadotrophic LH cells are restricted to the anterior third of the pars distalis. The gonadotrophic FSH cells, which are more common ventrally, contain large erythrophilic granules. The conical thyrotrophic cells are smaller and scattered throughout the gland. The alpha and X cells are indistinguishable from each other and are abundant posteriorly and medially; they are small cells with scanty cytoplasm and many moderate-sized erythrophilic granules. The size and shape of the nuclei of the various types of cells differ characteristically, but all are poor or very poor in chromatin.

In *Eryx jaculus* and *Lichanura roseofusca*, two semi-fossorial boids, the intermediate lobe is distinctly reduced and contains few chromophilic cells. The median eminence lies nearer the neural lobe and more dorsal to the anterior region of the pars distalis than in other boids.

Colubridae, Elapidae and Hydrophiidae

In members of these three families, the hypophysis is very similar to that of the non-burrowing Boidae, except that the median eminence lies farther anterior to the distal lobe to which it is joined by a thick connective tract and the pars tuberalis is better developed. Moreover, the pars distalis is slightly more elongate anteroposteriorly.

Cytologically, the distal lobe is characterized by the large size and extreme abundance of the gonadotrophic LH cells, which occupy most of the anterior two-thirds of the lobe and are filled with large, carminophilic, barely PAS-positive granules. The much less numerous gonadotrophic FSH cells are almost entirely restricted to the ventromedial and lateral regions and contain fine cyanophilic granules. The thyrotrophic cells, which are very abundant and scattered throughout the lobe, are also characteristic, being regularly conical with very large erythrophilic granules. The alpha cells are generally small and have few large secretory granules; they are numerous in the posterior third of the pars distalis. The slightly larger X cells are rare and restricted to the medial region; their granules are similar to those of the alpha cells, having most of the same staining properties but also reacting slightly to PAS.

In the Hydrophiidae, the hypophysis tends to be bilaterally symmetrical. The neural lobe, which is often richer in neurosecretory products than in the other families, lies dorsal or a little posterior to the distal lobe.

Aside from these very slight differences and a few small specific variations, the homogeneity of these three families is remarkable. However, in two colubrids from Madagascar (*Mimophis mahafalensis* and *Lycodryas* sp.) the morphology of the distal lobe and of its different types of cells is very similar to that in the Viperidae.

Viperidae

Anatomically, the hypophyses of the Viperidae (including the Crotalinae) differ from those of the preceding families only in the more massive form of the distal lobe, which is ovoid with a well-developed posterior region. This difference is greatly magnified by the relatively large size of the heads of viperids. Since a similar phenomenon occurs in the elapid *Acanthophis antarcticus*, it has no systematic value.

In contrast, the cytological characteristics of the different types of cells in the distal lobe are entirely distinct from those of the other Colubroidea, although their locations are identical. The less numerous gonadotrophic LH cells are amphophilic, strongly PAS-positive, and stain particularly deeply with alcian blue and paraldehyde fuchsin. The ovoid gonadotrophic FSH cells contain large secretory granules or small erythrophilic particles on a cyanophilic base. The thyrotrophic cells, which are fairly small and rather uncommon, are unusual in possessing many, quite large erythrophilic granules. The large, often cuboid alpha cells have large and generally very abundant secretory granules and hemispherical or flat basal nuclei; they occupy much of the enlarged posterior region. The X cells are much less numerous than the alpha cells and often form small homogeneous cords anteromedially. Their granules are distinctly less abundant than those of the alpha cells and slightly PAS-positive.

In the burrowing viper *Atractaspis* sp. the neural lobe lies distinctly lateral to the distal lobe. Moreover, the intermediate lobe is very reduced and formed entirely by involuted chromophobic cells. However, neither the form of the distal lobe nor its cytological characteristics are modified.

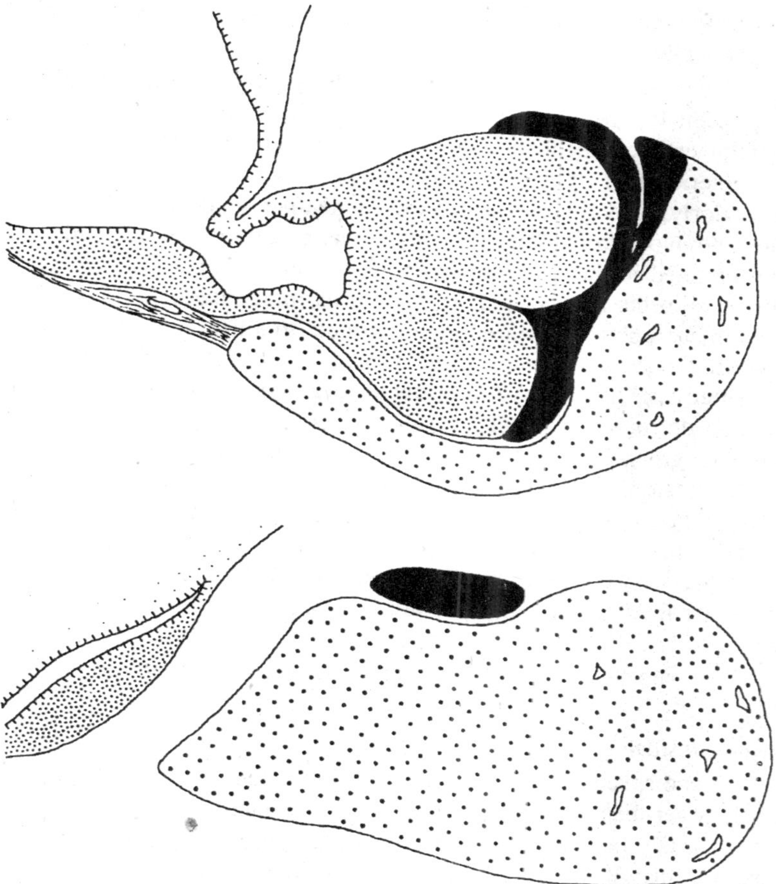

Fig. 2.10. Parasagittal sections through the hypophysis of Cerastes cerastes.

Testudines

Testudinidae

The ovoid neural lobe does not extend very far laterally and is more or less divided into hollow lobules. The infundibular recess penetrates it deeply, therefore the volume occupied by neurosecretory fibres is small. The median eminence is slightly swollen and lies anterodorsal to the anterior end of the distal lobe. The intermediate lobe is situated between the posterior half of the pars distalis and the neural lobe, which it surrounds laterally; it is little developed and thin medially where there may be a hypophyseal cleft, but is thicker laterally and is made up of irregular cords of small, weakly chromophilic cells.

The rather flat distal lobe overlaps the complex of neural and intermediate lobes, especially anteriorly. It is formed by irregular cellular cords, which often delimit a kind of thick cortical region in the anterior half of the lobe. There are very large vascular sinuses posteriorly, and all the cells show a distinct tendency towards an elongate ovoid form. The fairly numerous gonadotrophic LH cells are concentrated in the anterior region. The gonadotrophic FSH cells are abundant in the anterior half or two-thirds, particularly in the cortex. The thyrotrophic cells are often difficult to differentiate from the preceding ones and have the same distribution; however, their granules are often slightly larger and distinctly cyanophilic, whereas those of the gonadotrophic FSH cells are slightly amphophilic. The alpha cells are often grouped around homogenous pseudo-follicles and are abundant in the posterior region of the pars distalis, although they occur throughout all the lobe except its anterior end. They contain fine, very densely staining, orangophilic granules. The X cells, which are less numerous than the preceding ones, are especially concentrated in the anterior half; their fine granules are less dense than those of the alpha cells and are somewhat carminophilic.

The pars tuberalis, which is especially well developed in turtles, has already been described.

Pelomedusidae

In one specimen of *Pelomedusa subrufa*, the neural lobe appears much larger than in members of the preceding family. The infundibular recess hardly penetrates it, and the pars nervosa, which is somewhat similar to that of snakes, consists largely of solid, more or less spherical lobules between which pass the cellular cords of the intermediate lobe. The latter is distinctly better developed than in other turtles (18% of the hypophyseal volume, instead of 11-14%).

In contrast, the form of the distal lobe and the adenohypophyseal cytology show nothing remarkable. It is possible that the morphological characters described in *Pelomedusa* distinguish the pleurodires from the cryptodires; however, the examination of many other species would be necessary to confirm this.

Crocodilia

In *Crocodylus porosus*, *C. niloticus*, and *Alligator mississippiensis*, the neural lobe is small but rather massive; the flattened infundibular recess penetrates it deeply. The intermediate lobe is well developed

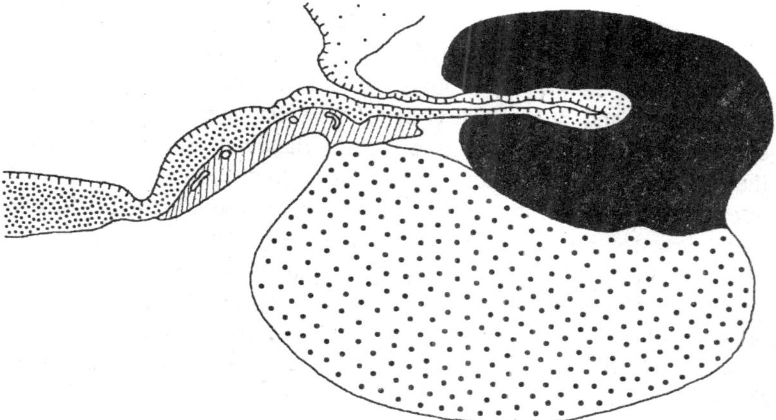

Fig. 2.11. Sagittal section through the hypophysis of Crocodylus niloticus.

and almost completely surrounds the neural lobe; however, the cellular cords are made up only of small, weakly chromophilic cells and some involuted cells. The distal lobe is massive and more or less ovoid. There are some small vascular sinuses posteriorly and a few undifferentiated cells in the thick cellular cords.

The adenohypophyseal cytology is only known in *Crocodylus niloticus*. The very numerous gonadotrophic LH cells are restricted to the anteroventral and lateral regions. Their very dense granules are erythrophilic and do not stain with hematoxylin lakes. The slightly larger gonadotrophic FSH cells are concentrated in the anteromedial region; they contain fairly fine, mostly amphophilic granules. The small, conical thyrotrophic cells are scattered throughout the distal lobe; they have much larger and distinctly basophilic secretory granules. The alpha cells are very small and have few orangophilic granules. They are restricted to the posterior third of the lobe. The numerous X cells arc carminophilic like the gonadotrophic LH cells, but have less dense secretory granules; they are most abundant in the dorsomedial part of the pars distalis.

In *Caiman crocodilus* the infundibular recess remains largely open, and the walls of the neural lobe are thinner than in the other Crocodilia studied. Moreover, the intermediate lobe is distinctly less developed (17% of the hypophyseal volume instead of 33%). In two very young specimens, the adenohypophyseal cytology differs markedly from that of *Crocodylus niloticus* of the same age, but only comparisons between adults could be used.

Conclusions

The morphology of the hypophysis is clearly modified in fossorial species. In all the very specialized families in which the representatives are limbless and more or less blind the sella turcica is only a slight depression, the hypophysis is distinctly flattened dorsoventrally with the neural lobe being lateral or posterior to the distal lobe, and the intermediate lobe is atrophied or absent. These characteristics are much less marked in burrowing snakes of relatively unspecialized families; first, there is a reduction in the size of the intermediate lobe as in Erycinae, and then that lobe atrophies as the neural lobe becomes clearly lateral (*Atractaspis*). These tendencies are not present in snake-like but non-burrowing lizards, such as *Anguis*, *Ophisaurus*, *Delma*, and certain *Chalcides*, nor in the arenicolous species that "swim" in the sand, such as *Scincus scincus* and *Uma inornata*. The families with the most specialized hypophyses are the Typhlopidae and Leptotyphlopidae, and then the Amphisbaenidae and Feyliniidae. The Trogonophidae and Anniellidae are distinctly less specialized.

Other modes of life do not seem to influence the morphology of the hypophysis. However, the posterodorsal (not dorsolateral) position of the neural lobe of the Hydrophiidae and the larger amount of neurosecretory products in this family should be noted. There are no significant differences between the terrestrial turtles (Testudininae) and those of fresh water (Emydinae). The intermediate lobe is well developed, even hypertrophied, in the lizards that change colour rapidly (Chamaeleonidae, *Anolis*, *Calotes*, *Phelsuma*). However, this lobe may be just as important in species in which the chromatophores undergo only slow and rather slight changes, such as many Agamidae, most Gekkonidae, and certain *Acanthodactylus*. In several groups, for example the Scincidae and Amphisbaenia, there is usually a positive correlation between the duration of daily exposure to the sun and the size of the intermediate lobe. However, there are numerous exceptions to this rule; in the Gekkonidae, for example, the intermediate lobe of diurnal species is no more developed than that of nocturnal ones.

Within broad limits, the morphology of the hypophysis can be correlated with the systematic position, and the complex of anatomical and cytological characters of this organ permits the recognition of several different structural types, which are homogeneous and well defined.

The chelonian type, studied in only a few species, is characterized principally by an elongated hypophysis, a very hollow pars nervosa (except in the only pleurodire examined), an intermediate lobe formed from irregular cellular cords, the presence of a well-developed pars tuberalis and of large vascular sinuses in the posterior region of the pars distalis, and finally the frequently lanceolate form of the adenohypophyseal cells.

The crocodilian type, also studied in only a few species, shows a massive, high hypophysis, a poorly developed neural lobe, and a large intermediate lobe made up of small, weakly chromophilic cells and rather specialized gonadotrophic LH cells.

The ophidian type (not found in typhlopids and leptotyphlopids) is far more distinctive than the chelonian and crocodilian types. The solid, spherical neural lobe lies dorsolateral to the pars distalis, so that the hypophysis is bilaterally symmetrical, as in all other non-burrowing reptiles. The pars terminalis is well developed, and the intermediate lobe is of a special type and is broadly united to the distal lobe. Three homogeneous groups may easily be recognized: (1) the Boidae, (2) the Colubridae, Elapidae, and Hydrophiidae, and (3) the Viperidae.

There appears to be a rhynchocephalian-saurian pattern characterized by a well-developed neural lobe which is deeply penetrated by the infundibular recess, a poorly differentiated pars terminalis, and an intermediate lobe which is always filled with large chromophilic cells, often arranged in two regular layers and separated by a distinct hypophyseal cleft. However, this group is not homogeneous, and contains several very distinct types.

The rhynchocephalian type has a poorly developed but distinct pars tuberalis and a massive pars distalis with very few, weakly chromophilic FSH gonadotrophic cells. The mode of contact between the neurosecretory fibres and the primary capillaries of the median eminence is especially characteristic and, indeed, unique among the studied vertebrates.

The gekkonid type, found also in the Pygopodidae and Xantusiidae, is very characteristic; the distal lobe has a well-developed anterior region and a flat posterior region with some scattered gonadotrophic LH cells. The intermediate lobe largely surrounds the neural lobe and possesses an internal layer formed from a single layer of large, regular prismatic cells.

The varanid type occurs in only one family. It differs from all the others in the general form of the massive and erect hypophysis and in the very special characteristics of most of the types of cells in the distal lobe.

The anguid and helodermatid type is easily recognized by the presence of very large vascular sinuses in the posterior region of the pars distalis and by the scarcity of undifferentiated cells. All the cells of the adenohypophysis are large, particularly the very numerous gonadotrophic LH cells; in contrast, and exceptionally in the Sauria, the X cells are rare and smaller than the alpha cells, and have the same staining properties as the latter. The numerous cuboid thyrotrophic cells are regularly arranged along the vascular sinuses. The nuclei are particularly large, spherical, and clear.

The scincid type is very homogeneous and is represented in a single family. It shows some common points with the preceding types: the small number of undifferentiated cells, the abundance and large size of the gonadotrophic LH cells, and the form and distribution of thyrotrophic cells. It also resembles the lacertid type in having numerous, small, non-granular alpha cells and large carminophilic X cells.

The iguanid-lacertid type is, unlike those just reviewed, not very homogeneous. The anatomy is characteristic, with the internal layer of the intermediate lobe being formed of stratified or irregular, pseudo-stratified cells, the small alpha cells being abundant and more or less nongranular, the X cells being large and carminophilic, and the gonadotrophic LH cells being rare and, in the Iguania, weakly chromophilic. The intermediate lobe is always hypertrophied in the Chamaeleonidae, the Agamidae, and Iguanidae of the genus *Anolis*.

It is clear that the limbless and blind fossorial squamates of different families are not closely related despite their numerous convergent characteristics. The cytological characteristics of the pars distalis, which do not resemble modifications of a particular type for subterranean life except for a distinct tendency to the reduction in size of the cells, differ greatly even between the Amphisbaenidae and the Trogonophidae. On the other hand, these characteristics clearly link the Anniellidae with the Anguidae and Helodermatidae, and, to a lesser extent, the Feyliniidae with the Scincidae. The anatomical resemblance of the hypophyses of the Typhlopidae and Amphisbaenidae is evidently the result of convergence. The Leptotyphlopidae, which have reduced neural lobe, a hypertrophied median eminence, and some very small cells with little secretory product, are quite different.

In most respects, the study of the hypophysis is consistent with the classification adopted by Romer (1956). However, some points require further discussion. The morphology of the hypophysis suggests a very isolated position for the Varanidae and, on the other hand, close relationships between the Anguidae and Helodermatidae and among the Gekkonidae, Xantusiidae, and Pygopodidae. The homogeneity of the Colubridae, Elapidae, and Hydrophiidae and of the Viperinae and Crotalinae is noteworthy.

No single criterion suffices to establish or modify a classification. If the observations on the comparative morphology of the hypophysis are confirmed on additional characters, they would suggest the following modifications of Romer's classification:

1. Inclusion of the Xantusiidae in the Gekkota.
2. Recognition of the Feyliniidae as a distinct family, closely related to but not included in the Scincidae.
3. Differentiation of the Scincoidea from the Lacertoidea if the infra-order Scincomorpha is to be retained as a unit. As noted before, the hypophysis shows some resemblances among the Scincidae, Anguidae, and Helodermatidae on one hand and between the Lacertoidea and Iguanidae on the other.
4. Placement of the Helodermatidae in the super-family Anguoidea rather than the Varanoidea.
5. Recognition of the Trogonophidae as a family, related to but distinct from the Amphisbaenidae.
6. Elevation of the Typhlopidae and the Leptotyphlopidae to the rank of distinct infra-orders.
7. Placement of the Booidea and the Colubroidea in the same infra-order.
8. Recognition within the Colubroidea of two distinct lineages, one including the Colubridae, Elapidae, and Hydrophiidae, and the other the Viperidae (including the Crotalinae). If it were not for the taxonomic problems caused by the large number of colubrids, one would reduce these four families to two.

These suggestions have been made only to summarize the results that can be derived from the various studies of the reptilian hypophysis. It must again be emphasized that these observations, based on one organ, have only limited weight. It would be just as regrettable to have a classification based solely on this character as to ignore it completely.

The study of the comparative morphology of the reptilian hypophysis is not complete. This organ has not been examined in numerous families, such as the Gavialidae, many Testudines, some limbless burrowing lizards of doubtful affinities (Dibamidae, Anelytropsidae), the Xenosauridae (including *Shinisaurus* of unclear systematic position), the Lanthanotidae, and all the more or less fossorial families related to the Boidae. It would also be interesting to study the hypophysis in some already fairly specialized burrowing lizards which belong to families, such as the Scincidae and Teiidae, members of which live above ground; this has been done in the case of certain groups of snakes. Finally, the significant differences which exist between the two main lines of the Colubroidea might allow the recognition, if they still exist, of some colubrids which have retained certain primitive characteristics and which might have given rise to the Viperidae.

It is thus clear that the morphology of the reptilian hypophysis shows important variation between the different families or orders and also sometimes depending on the mode of life. However, if the latter factor and several exceptions are ignored, it is possible to separate the essential characters and to compare them with those of other tetrapods. The presence of a well-developed pars tuberalis (except in the Squamata; however, in this case, regression appears to be secondary) and the markedly pediculate structure of the neural lobe, which is separated from the infundibulum by a well differentiated pituitary stalk, distinguish reptiles from amphibians and show a closer resemblance to conditions in birds and mammals. The very clear zonation of the pars distalis is a common characteristic of reptiles and birds, but the latter, like certain mammals, lack an intermediate lobe. It should be noted that no other vertebrates show as marked hypertrophy of the intermediate lobe as do many reptiles. Finally, the morphology and histochemical characteristics of the different types of reptilian adenohypophyseal cells resemble more closely those of birds and mammals than those of amphibians.

3

THYROID GLAND

Studies on a wide variety of vertebrates have clearly demonstrated that both the morphology of the functional unit of the thyroid gland and the mechanism of production of the thyroid secretion are essentially identical throughout the group. In all vertebrates the gland is composed of more or less spherical follicles, each made up of epithelial cells arranged in a single layer around a lumen filled with secreted material, the thyroid "colloid". In most vertebrates the follicles are closely associated to form a well-defined compact gland in which the individual units are bound together with loose connective tissue, the whole enclosed in a connective tissue capsule. However, cyclostomes and most teleost fishes have the follicles diffusely spread throughout the tissue beneath the pharynx. The gland is always well vascularized, usually with extensive networks of capillaries around the follicles. The height of the follicular epithelium, the amount of colloid present, the staining reaction of the colloid, and the extent of vascularization all vary with the degree of functional activity of the gland. It is therefore possible to utilize histological criteria as one, largely qualitative, indication of the level of secretory activity of the thyroid.

The outstanding physiological feature of the thyroid is its ability to concentrate iodine from the circulation. The biochemical mechanisms by which this is accomplished have been studied intensively. Briefly, the following steps, most of them enzymatically controlled, are involved. Iodine, in the form of iodide, enters the blood stream from the food and also, in the case of fishes and amphibians, from the surrounding water. Iodide is concentrated from the blood by the cells of the thyroid follicles ("iodide pump" of the thyroid). It is then oxidized and organically bound. The latter involves the formation of monoiodotyrosine (MIT), diiodotyrosine (DIT), tetraiodothyronine (thyroxine, T_4), and triiodothyronine (T_3). All these compounds are present in bound form in the thyroid protein, thyroglobulin, which accumulates in the colloid. The biosynthesis of thyroglobulin probably occurs in the intrafollicular colloid rather than in the cells themselves. The binding of iodine in this way makes possible the concentration of this element in the thyroid even when the rate of intake into the body is extremely low. Thyroglobulin may be stored indefinitely in the colloid or it may at any time undergo hydrolysis to yield free MIT, DIT, T_3, and T_4. The latter two compounds, which are regarded as the active hormones of the thyroid, are then able to pass into the blood stream and reach the general body tissues. Most of the thyroid hormone found in the circulating blood is in the form of T_4 which does not enter the red blood cells in any significant concentration but is bound to plasma proteins. Free MIT and DIT undergo deiodination in the colloid, and the iodide may then return to the circulation to be again picked up by the thyroid cells or to be eliminated from the body. These various steps may occur rapidly or slowly depending upon environmental conditions and upon the physiological state of the animal. Therefore, some of the most useful indices of thyroid activity are measurements of the rate and amount of iodide uptake, the rate of iodide turnover, and the rate of incorporation of iodine into the various compounds involved in synthesis of the hormone. Such measurements have been made feasible primarily through the use of radioisotopic techniques and chromatographic analysis.

The thyroid is one of the endocrine glands whose functioning is controlled by a specific hormone of the pituitary. The thyrotrophic or thyroid-stimulating hormone (TSH) is a glycoprotein secreted by certain basophilic cells of the anterior lobe of the pituitary. Administration of TSH causes an increase in thyroid activity that may easily be demonstrated by both histological and biochemical methods. Absence of TSH, which can be achieved by hypophysectomy, results in cessation of thyroid activity and atrophy of the gland. In a normal animal the thyroid and pituitary exert reciprocal effects upon each other by means of a feedback mechanism that results in maintenance of the normal balance of hormone production.

Several reviews concerning thyroid morphology and physiology have dealt with reptiles in greater or less detail. For this reason much of the earlier work will be only briefly summarized in the present account.

Gross Morphology

In turtles and in snakes the thyroid is an impaired, roughly spherical gland, lying ventral to the trachea and just anterior to the heart. In *Sphenodon* it is a single narrow body, transversely elongate, in the same general position. Members of the Crocodilia have a thyroid with well-defined lobes on the two sides of the trachea connected by a narrow isthmus. Lizards exhibit a wide variety of thyroid forms with unpaired, bilobed, and completely paired glands found even in different members of the same family.

Detailed descriptions of the anatomical relations of the thyroid are available for only a small number of reptiles. In the turtle *Emys orbicularis* the gland is a spheroidal structure lying in the cavity of the arch of the innominate trunk, just anterior to the heart and ventral to the point where the trachea divides into the two bronchi. Its blood supply is through two pairs of arteries. The superior thyroid arteries are branches of the carotids. They are rather thin, variable in position, and sometimes absent. The inferior thyroid arteries are larger, short, vessels that emerge directly from the innominate trunk just below the point where it gives rise to the right and left carotid and subclavian arteries. The thyroid veins join the accessory pectoral veins and empty into the subclavians at the junction of the jugular and axillary veins. A lymphatic network is found around the external surface of the thyroid and this connects with the lymphatics of the neck. Innervation is derived from the laryngeal branches of the vagus and from fine branches of the cervical sympathetics which accompany the arterial supply into the gland. In this species the thyroid may be exposed by trepanning if one applies the instrument one-half centimeter anterior to the point where hyoplastral and hypoplastral plates join. If the bone is removed and the forelimbs are drawn apart so as to put tension on the scapuloclavicular ligaments the gland can be visualized by extending the animal's neck. Testudo graeca shows essentially the same features.

Published accounts of thyroid morphology in snakes are limited to brief descriptions for *Natrix*, *Coluber*, and *Malpolon*. *Natrix sipedon fasciata* was reported by Thompson (1910) to have a median, spherical gland, and Bragdon (1953) stated that *Natrix sipedon sipedon* possesses an ovoid or globular thyroid, measuring 3.0 by 4.0 mm in a large female, and located at the level of the twenty-first to twenty-fourth ventral scute. Francescon (1929) referred briefly to the thyroid in *Coluber viridiflavus* and *Malpolon monspessulanus* in connection with his observations on the ultimobranchial body, but gave neither figures nor detailed descriptions.

The thyroid of *Sphenodon punctatus* was described by van Bemmelen (1887) as a single transversely elongate body lying across the trachea anterior to the heart. O'Donoghue's (1920) account of the blood vascular system of *Sphenodon* includes a figure showing these features and a description of the superior thyroid and inferior thyroid arteries which branch respectively, from the external carotid artery and the pulmonary arch. The paired thyroid veins reach the precaval by way of the tracheal veins. The description given by Adams (1939) agrees closely with these accounts.

The lizard and amphisbaenian thyroid is of special interest. As early as 1844, Simon pointed out that the general form of the gland differs widely in representatives of different families of lizards so that, whereas the members of some families have an unpaired median thyroid, other lizards have completely paired lateral thyroids, and still others have a bilobed gland with a narrow connecting isthmus. It is now clear that in only a few families are all members of the family uniform with respect to thyroid morphology and, in fact, several of the larger families have representatives showing each of the three types mentioned. On the other hand, a particular thyroid form is always characteristic for a given species and usually for the genus as well. Thyroid morphology may, therefore, be of some significance in indicating systematic relationships in lizards. The accompanying table summarizes presently available data, some of it not previously published, on the form of the thyroid gland in the various lacertilian and amphisbaenian families. Earlier references on this subject may be found in the review paper of Lynn (1960).

It may be noted that, if one considers all lizards but not the members of the Amphisbaenia, the most common thyroid form is like that of mammals, a bilobed gland with a connecting isthmus beneath the trachea. A single broad thyroid stretching across the trachea is found almost as frequently however, and these two types often occur in different genera of the same family. Single, rounded thyroids are rare and appear to be characteristic of only two small families, Pygopodidae and the monogeneric Anniellidae. It is noteworthy that completely paired thyroids, aside from their occurrence in two closely related monogeneric groups, are found in only four other families all of which also have representatives with bilobed and single broad glands. Among these four families two (Iguanidae and Agamidae) belong to the Ascalabota and the other two (Teiidae and Anguidae) belong to the

Autarchoglossa. These facts may be taken to indicate that the paired condition has evolved independently at least twice in lizards and that pairing of the thyroid may be regarded as derived from the unpaired condition, probably with bilobing as an intermediate step.

The amphisbaenians prove to be aberrant in thyroid form as they are in so may other features. A bilobed thyroid is rare in this group. Paired glands are common but are of two diverse types. Although it is not indicated in the table, some species have paired thyroids of the usual spherical or ovoid shape, but others have extremely elongated, attenuated, thread-like glands which, though less than 1.0 mm wide, stretch alongside the trachea for as much as several centimeters. Such thin elongate thyroids are not known to occur in other reptiles or, indeed, in any other group of vertebrates. In the family Trogonophidae most genera have single thyroids of elongate form and these are sometimes found to one side of the trachea rather than in the midline. In this family only the genus *Trogonophis* has paired thyroids, but these are elongate and are also connected by an isthmus at the posterior end. Finally, the position of the thyroid is unusual in amphisbaenians. In all cases the gland lies much further forward than in other reptiles. It is, therefore, not situated near the heart but lies either midway along the trachea or near its anterior end. Thyroid morphology does not support any concept of a close relationship between the Amphisbaenia and the snakes; in no amphisbaenian does the form or position of the thyroid resemble that of snakes.

Histology and Cytology

The reptilian thyroid is always enclosed in a capsule of connective tissue containing argyrophilic fibers and, in some cases, a few elastic fibers. Scattered melanophores may also be found in the capsule. Connective tissue septa can sometimes be traced from the capsule into the gland dividing it into ill-defined lobules, and a delicate areolar connective tissue network containing argyrophilic fibers always surrounds the follicles. The follicles vary in diameter from 50 μ to 300 μ and may reach much larger sizes in large turtles. They are commonly rounded but may be of quite irregular shape, particularly in glands that are in phases of intense secretory activity. The simple epithelium lining the follicles varies in height from flat to columnar depending upon the functional state of the follicles and the amount of colloid present in its lumen. Mitotic figures may be seen in the epithelium during periods of increased thyroid activity. The colloid masses within follicles are eosinophilic. They may have a uniform homogeneous appearance or may exhibit large numbers of chromophobe droplets, especially at their peripheries. The epithelial cells have basally situated nuclei, usually with rather indistinct nucleoli. The cytoplasm may contain well-defined fuchsinophilic granules which are taken to be droplets of secretion.

Early studies of the normal histology of the thyroid in a variety of reptiles are cited by Lynn (1960). They include Viguier's (1909a, b, 1911c) brief descriptions for several lizards (*Tarentola mauritanica*, *Lacerta lepida*, *Psammodromus algirus*, and *Chalcides ocellatus*), Naccarati's (1922) accounts for *Emys orbicularis* and *Testudo graeca*, and Hellbaum's (1936) study on *Thamnophis radix* and *T. sirtalis*. Normal thyroid histology is also discussed in a number of papers that are primarily concerned with the effects of various experimental treatments. The earliest of these is the study of Galeotti (1897) concerning changes in the follicular epithelium of the thyroid in turtles treated with certain toxic substances. Barchiesi's (1928) observations on the effects of prolonged starvation on the thyroid of *Emys orbicularis* are noteworthy in that they emphasize the high degree of individual variation in histological appearance of the gland in normal turtles. An important source of information on the histology of the reptilian thyroid is to be found in the numerous studies of seasonal variation in the gland and in studies on the relation between thyroid activity and molting.

Several recent studies deserve special mention. Yamamoto (1960) has provided further histological data for the following species: *Clemmys japonica*, *Lepidochelys olivacea*, *Gekko japonicus*, *Eumeces l. latiscutatus*, *Elaphe climacophora*, *E. quadrivirgata*, *E. conspicillata*, *Natrix v. vibikari*, *N. t. tigrina*, and *Agkistrodon halys blomhoffii*. The general histological features in these reptiles are in accord with those summarized above. In addition Yamamoto found a sparse scattering of epithelial cells which show the characteristics of the "colloid cells" of earlier authors. They stain red-purple with PAS and have small, sometimes pycnotic, nuclei. They are interpreted as either degenerating or resting cells. He also reported that some of the snakes he examined had funnel-shaped sphincter apparatuses at the branching points of arteries in the thyroid which resemble such sphincters found in mammals. Certain spherical or ovoid cysts consisting of a stratified epithelium enclosing degenerating cells, cellular detritus, and a crystalloid material were sometimes found in the thyroids of *Clemmys japonica* and *Elaphe climacophora*. Yamamoto considers that these represent similar enigmatic structures of mammals which may be of branchiogenous origin.

Large ovoid cells known as "macrothyreocytes" or "parafollicular cells" have been found in the thyroids of mammals, and these also occur in reptiles. They are usually intra- or subepithelial rather than truly parafollicular and they enclose a secretory product in the form of PAS-positive granules or large droplets. The secretion is reported to contain muco- or glycoproteins as well as amino acids and protein-bound sulfhydryl groups. In snakes (*Python*, *Vipera*, *Cerastes*) the secretory granules coalesce to form large homogeneous masses which may occupy the greater part of the cell. Macrothyreocytes increase in number during thyroid activation and intermediate forms between follicle cells and macrothyreocytes are not uncommon. Kroon suggests that the macrothyreocyte may represent a phase in the development of ordinary cells of the follicular epithelium and that their morphological features may be interpreted as evidence of intracellular colloidogenesis.

Blood cells are occasionally found in the colloid. Intrafollicular granulocytes have been reported in *Lacerta* and *Chelydra*. Miller (1955) found that both eosinophilic granulocytes and erythrocytes occur in the colloid of *Xantusia* following any period of thyroid hyperactivity when there is a sudden thinning of the epithelium.

There are relatively few studies of the ultrastructure of the reptilian thyroid. Kurosumi (1961) presents a photomicrograph of an electron microscope preparation of part of a thyroid cell of the snake *Elaphe quadrivirgata* which shows a well developed endoplasmic reticulum with unusually large cisternae. Secretory granules of highly variable size and density are present, and the smallest of these are said to resemble Golgi vesicles. The nuclear envelope has well-defined pores.

With advancing age the histology of the lizard thyroid is altered by gradual invasion of lymphocytes which spread throughout the tissue or may form large nodules. Some follicles undergo involution and fatty tissue appears in the resulting spaces. Brown pigment granules in the supranuclear zone in some snakes have also been interpreted as characteristic of advanced age.

Embryology

There is no indication that reptiles are in any way unusual in the mode of origin and development of the thyroid. Studies have been made on representatives of all the major groups of living reptiles except the Rhynchocephalia. In all of these the gland has been shown to arise as a midventra outpocketing of the embryonic digestive tract at the level of the first pair of pharyngeal pouches. A connection with the pharynx (thyreoglossal duct) persists only briefly and the tissue mass elongates posteriorly in the form of solid strands or cords. These then break into groups of cells that begin to form primary follicles. The lumina of newly organized follicles usually contain only small amounts of a non-stainable colloid, but the typical staining reaction appears within a short time. Meanwhile the follicles gradually migrate posteriorly to their definitive position and become encapsulated to form a compact gland.

Raynaud and Raynaud (1961) and Raynaud et al. (1963) report that a number of endocrine organs (thyroid, gonads, suprarenals, hypophysis) exhibit indications of precocious secretory activity in *Anguis fragilis*. Cells of the thyroid Anlage are found to contain fine PAS-positive granules in their cytoplasm at early stages before the appearance of organized cell cords or follicles. This material is later seen in intercellular spaces and still later (embryos of 21 or 22 mm crown-cloaca length) can be found at the centres of cell cords. In the viviparous lizard *Xantusia vigilis*, organized follicles containing thin colloid are present at the 7 to 8 mm stage which is about one week before the basophil cells of the anterior pituitary are differentiated. The embryonic thyroid shows histological evidence of high activity during the last third of the gestation period.

Dimond's (1963) investigation of thyroid embryology in four species of turtles shows that organized follicles are present during the second quarter of development and are capable of binding iodine. Thyroid hypertrophy and hyperplasia results when potassium perchlorate or thiourea are administered to embryos shortly after the mid-stage in development, and this is taken to indicate that pituitary stimulation of the thyroid has begun by this time. Pituitary beta cells, the presumed source of TSH, are also demonstrable at this stage.

Thyroid-pituitary Relations

Effects of Thyroidectomy, Hypophysectomy, T_4 Administration, or TSH Administration

The reptilian thyroid, like that of other vertebrates, is under pituitary control. Evidence for a relationship between thyroid and pituitary function was first obtained by studies of histological changes in the pituitary in thyroidectomized animals. Viguier (1911b) reported an increase in basophil cells of the anterior pituitary in lizards (*Uromastyx*) killed 8 to 10 weeks after having been thyroidectomized and similar results have been obtained for *Thamnophis radix* and *Testudo graeca*. Recent interest in the histology and cytology of the pituitary gland and in neurosecretory activity in the hypothalamus has led to histochemical characterization of pituitary cell types in a number of reptiles and to some definition of neurosecretory pathways in reptiles. It is to be expected that these studies will be followed by observations on thyroidectomized animals which will lead to a more precise identification of the cells of the anterior pituitary which are responsible for the secretion of the thyroid-stimulating hormone (TSH) in reptiles and to a better understanding of the mechanism of action of the thyroid-pituitary axis.

Histological observations concerning seasonal changes in the pituitary which can be correlated with the thyroid cycle also indicate dependence of one upon the other. Some observations of this kind have been made on *Xantusia vigilis* and on *Clemmys caspica leprosa*.

Removal of the hypophysis results in repressive changes in the thyroid characterized by flattening of the follicular epithelium and failure of release of the colloid. Hellbaum found that daily administration of anterior pituitary extract to snakes which had been hypophysectomized earlier caused a rapid return to normal thyroid histology. Hypophysectomy also brings about a marked reduction in the per cent uptake of radioiodine by the thyroid.

Administration of pituitary extracts or of purified TSH preparations to unoperated animals results in histological evidence of increased activity in the thyroid although the degree of response shows much individual variation. Eggert was unable to demonstrate any thyroid response to TSH in lizards treated during hibernation. Turtles (*Chrysemys*) given 36 daily injections of pituitary extract showed a thyroid response that persisted as long as 136 days after the first injection. *Natrix* (= *Tropidonotus* auct.) *natrix* gives a well defined and very constant histological response to TSH even at temperatures as low as 13°C.

As would be expected, administration of exogenous T_4 to an animal results in a reduction of the activity of its own thyroid since a high level of thyroxin in the blood causes a decrease in production of TSH by the pituitary. This has been demonstrated for two lizards (*Lacerta*, *Gekko*) and one snake (*Chionactis*).

Effects of Goitrogenic Drugs

There are various drugs that inhibit the production of thyroid hormone either by interfering with the uptake of iodide or by inhibiting the formation of iodinated protein. Administration of these substances can thus cause a functional thyroid deficiency which may be severe enough to correspond closely to that resulting from surgical thyroidectomy. Animals so treated show a thyroid response characterized by hyperemia and by hyperplasia and hypertrophy of the follicular epithelium. If treatment is long-continued the whole gland may enlarge strikingly to form a definite goitre. These responses are attributable to a high output of TSH by the pituitary which has been stimulated to excessive thyrotrophic activity because of the low level of thyroxin in the blood. The use of goitrogens thus provides a less radical method than thyroidectomy for studying the pituitary's response to thyroid deficiency. Various turtles, both embryonic and adult, have been subjected to such treatment. The changes in thyroid histology are always clear-cut although there is much individual variability in the degree of responsiveness. The thyroids of most lizards studied show a well-marked histological response to goitrogen administration, but *Lacerta* is reported to be very unreactive.

The effects of goitrogens upon the uptake of radioiodine by the thyroid have been investigated in *Chrysemys*, *Pseudemys*, *Anolis*, *Xantusia*, and *Alligator*.

Biosynthesis of Thyroid Hormones

The basic similarity in thyroid physiology in reptiles and other vertebrates was indicated by early studies showing that the reptilian thyroid has a high iodine content, a certain proportion of which is present in thyroxin, and that feeding of reptilian thyroid is effective in inducing precocious metamorphosis in amphibians or in raising the basic metabolic rate in rabbits.

Recent work using radioiodine and chromatographic analysis serves to elucidate some of the details of biosynthesis of thyroid hormones in reptiles. In both *Terrapene carolina*, a terrestrial turtle, and *Pseudemys floridana*, an aquatic species, the thyroid is capable of accumulating I^{131} and synthesizing T_4, and it responds to TSH injection by an increased rate of I^{131} accumulation. Turtles kept under wet conditions show less I^{131} accumulation in the thyroid and more I^{131} excretion in the urine than do those kept under dry conditions. Under dry conditions as much as 80%, of the administered radioiodine is present in the thyroid by 8 days after administration. There is almost no excretion of I^{131} in the urine under these conditions, and it is postulated that, due to urinary retention, the urinary bladder acts as a reservoir for radioiodine which is slowly returned to the circulation by way of the cloaca. There is also a seasonal difference in thyroid function in these turtles. *Pseudemys* studied in March at one to four days after I^{131} administration showed incorporation in MIT and DIT only. Similar experiments in September revealed the presence of labelled T_4, as well as MIT and DIT, within three days. The amount of labelled T_4 then increased throughout the 27-day period of study. T_3 was also found, although in very small amounts.

The pattern of uptake and release of I^{131} by the thyroid in *Anolis carolinensis* and *Sceloporus occidentalis* has also been studied for both normal and dehydrated animals. In these animals the uptake of I^{131} by the gland did not differ in the two groups. The maximum uptake did not occur until 24 to 48 hours after administration, but thereafter there was no decrease in level for as long as eight days. This plateau level was about 10% of the injected dose in *Anolis* and 15% in *Sceloporus*. These observations suggest a rather slow thyroidal accumulation of I^{131} from the blood and reservoir tissues and a low rate of turnover. Synthesis of thyroid hormone also occurred slowly. MIT and DIT were found at highest levels at 24 to 48 hours after I^{131} administration and these decreased progressively while the levels of T_3 and T_4 (plus two unidentified radioiodinated compounds) gradually increased. In contrast to the results of the experiments on turtles, these studies show no well-defined effect of dehydration on the metabolism of radioiodine.

The uptake and release of I^{131} by the thyroids of lizards (*Anolis carolinensis*) acclimated at two widely differing temperatures (15° and 35°C) show marked differences. At 15° the highest levels of radioactivity are found during the first 24 hours after I^{131} administration and there is then a gradual decline throughout the next five days. At 35° there is a rapid increase in I^{131} accumulation to reach a peak which is much higher than is ever attained at 15°. This peak occurs at 48 hours and is followed by a fairly rapid decrease in activity so that by six days after injection the level is the same as that found in the lizards kept at 15°. Thiourea administration to animals maintained at these two temperatures strongly depresses the initial uptake of I^{131} in the thyroid, but specimens kept at 15° show a slower rate of turnover than do those kept at 35°.

Xantusia, a nocturnal lizard, kept at 26°C, shows the expected responses to thiourea or TSH treatment with respect to radioiodine uptake and release, but also gives some evidence of diurnal fluctuations in thyroid activity that may be correlated with its nocturnal habits.

In young alligators I^{131} uptake by the thyroid rises steadily to reach a peak of 19.6% of the injected dose at five days, and this declines to 7.8% by seven days. TSH administration causes only a slight loss of radioactivity; propylthiouracil or KI injection reduces the uptake sharply. Young turtles (*Pseudemys scripta*) reach peak uptakes earlier and at somewhat higher levels and also show a marked response to propylthiouracil, KI, or T_4 injection. Late embryos show relatively lower levels of uptake but also respond to propylthiouracil.

Differences in the rate of accumulation of I^{131} by the thyroid of *Lacerta* as compared with that of the dogfish have been interpreted as indicating differences between, the active agents of the thyroid in these two animals.

SEASONAL CHANGES IN THE THYROID

Seasonal Changes Related to Temperature

Seasonal changes in the histological appearance of the thyroid in temperate zone lizards were reported by Weigmann (1932) for *Lacerta*. These changes, as might be expected, indicate a higher functional activity in summer than in winter. Eggert's (1935) study for three different species of *Lacerta* confirms and extends Weigmann's observations. Moreover, his investigation of thyroid histology in lizards kept in the laboratory under controlled temperature conditions leads to the conclusion that the seasonal changes seen in animals collected in the field are mainly dependent upon environmental temperature. Gravid females of *Lacerta vivipara* have thyroid levels slightly higher than those of non-gravid animals, but Eggert considers that this is also a temperature effect related to the fact that pregnant lizards expose themselves to the sun for longer periods than usual.

Sceloporus undulatus shows a similar seasonal variation in thyroid histology, but in addition to temperature-dependent changes there are alterations in the thyroid connected with breeding activity. Wilhoft's work also clearly emphasizes the fact that observations on thyroid histology in lizards collected in the field may be quite misleading in indicating a direct relation between air temperature and thyroid activity. Because of the well-known basking habit of reptiles, records of average daily air temperature may differ widely from the temperatures to which the animals are actually exposed. When the air temperature is low most reptiles spend a large proportion of the daylight hours basking in direct sunlight, but periods of extremely high air temperature may be spent entirely in shaded areas at significantly lower temperatures.

Wilhoft found that thyroid activity, as judged by histological criteria, in *Sceloporus* fell steadily from July into September despite the fact that air temperatures in the collecting area were steadily rising during that time. Field observations showed that the animals were less active during this period and some could be said to be aestivating. Experimental exposure of *Sceloporus* to continuous high temperatures (34°-35°C) for periods up to thirteen weeks caused marked increases in thyroid epithelial height beginning within three weeks. A high mortality occurred in animals given the high temperature treatment for longer than six weeks.

Lizards that inhabit warmer climates and do not have a hibernating period give histological signs of highest thyroid activity during the cool winter months, possibly in relation to the maintenance of feeding activity in the winter, and a decrease in physical activity in the summer. *Leiolopisma rhomboidalis* of tropical Australia shows less seasonal variation in thyroid epithelial height than do either *Sceloporus* or *Xantusia*, but, in general, the lowest activity occurs during the cold season (May to August). In this species there are also variations in thyroid activity in both males and females that are related to the reproductive cycle. In *Agama agama savattieri* a marked increase in thyroid activity is found in March and April as compared with January and February. This apparently corresponds with a notable change in humidity rather than an increase in environmental temperature. It is probably also related to an increase in physical activity at this time and to the onset of the breeding season. Young *Anguis fragilis* are reported to show the highest thyroid activity in April and May.

Few studies of seasonal variations in the thyroid are available for snakes. Both *Vipera aspis* and *V. berus* exhibit low levels of thyroid activity during hibernation and higher levels in the summer, but the differences are more marked in *V. berus*, specimens of which were collected in the *Alps* at altitudes of 3,300 to 5,600 feet, than in *V. aspis*, a lowland species. In both, however, the highest levels of thyroid activity are associated with reproductive events. A similar situation is reported for *Natrix maura*.

Turtles (*Chrysemys picta bellii*) exhibit evidence of very low thyroid activity during hibernation. There is a direct relation between environmental temperature and the per cent uptake of radioiodine by the thyroid of *Pseudemys* and *Terrapene*. Moreover, injection of TSH is unable to influence the rate of thyroidal accumulation of I^{131} at very low temperatures (2°-3°C).

The thyroids of lizards (*Anolis carolinensis*) maintained in the laboratory at 35°C show histological evidence of slightly higher activity and much less variability than do those of animals kept at 15°C, but the effect of this temperature difference upon the functioning of the gland, as indicated by its uptake and release of radioiodine, is very great. We have obtained similar results for *Phrynosoma cornutum*.

The length of day may also influence thyroid activity. Exposure of turtles (*Testudo horsfieldii*) to continuous illumination causes a stimulation of the thyroid which reaches a peak within five days but then gradually declines so that the gland becomes inactive by the thirty-fifth day.

Seasonal Changes Related to Reproductive Cycles

Eggert reported that he could find no evidence of histological changes in the thyroid correlated with the breeding season in *Lacerta*. However, he found that thyroidectomy results in striking effects on both gonadal structure and breeding behaviour. The gonads show-degenerative changes which ultimately result in loss of reproductive ability. Males fail to get the spring nuptial coat and do not mate. Eggert postulated that the effect of thyroidectomy in the gonads may be indirect, partly due to a decrease in general metabolism and partly to some interference with anterior pituitary function. Wilhoft's (1958) measurements of the height of the thyroid epithelium in adult *Sceloporus occidentalis* collected throughout the period of activity shows a gradual increase in epithelial height in both sexes through the period of mating in April or May until June. Thereafter there is a rapid decrease in females immediately after egg-laying. Males show a more gradual decrease beginning in July, and, as has been noted above, these decreases continue into September despite the fact that environmental temperatures are increasing during this period. Study of the gonads indicates that spermatogenesis occurs during or shortly after emergence from hibernation and thus takes place during a time of relatively low thyroid activity. There is also a period of increase in testis size in September when thyroid epithelial height is again decreasing. In the females there seems to be a definite peak in thyroid activity associated with yolk deposition and ovulation. In *Anolis carolinensis* there is an increase in thyroid epithelial height in ovulating females, and experiments indicate that an increased level of ovarian hormones may be responsible for the thyroid response.

The Japanese lacertid lizard *Takydromus* exhibits characteristic changes in the Golgi apparatus of the cells of the thyroid epithelium during the breeding season, and these changes can be induced in castrated animals by the injection of sex hormones. The thyroid also shows a cycle that is related to the reproductive system in *Anguis fragilis*, *Xantusia vigilis*, a viviparous lizard, has a gestation period of three months. The thyroid of the female is hyperactive during the period of yolk deposition, ovulation, and the early part of gestation, declines in activity during the latter half of gestation, and is relatively inactive by the time of parturition. The male thyroid, though less markedly hyperactive than that of the female, shows increased activity correlated with spermatogenesis, development of sex accessories, and copulatory behaviour in the spring and another peak of activity in the fall when a new cycle of spermatogenesis begins. Wilhoft's (1963, 1964) findings for *Leiolopisma rhomboidalis* also show high thyroid activity in females during yolk deposition and in males during spring and fall periods of spermatogenesis and increased reproductive activity.

The sexual cycle of *Vipera aspis* is characterized by two periods of mating activity each year. Copulation occurs shortly after emergence from hibernation (March-April) and again in the fall (late September). Females have a two-year cycle. The first year, following the spring mating, yolk deposition and ovulation occur and gestation begins. The second year there is no spring mating but copulation does occur in the fall. Males show a slight rise in thyroid activity during the two mating periods, but no change that can be correlated directly with spermatogenesis. Females show similar increases in spring and fall, but the thyroid epithelium is at a low level during both yolk deposition and gestation.

In the cobra (*Naja naja*) there is a peak of thyroid activity in males just after emergence from hibernation in the spring which corresponds with the time of active spermiogenesis and mating. Moreover, a secondary secretory peak occurs during the period of testicular recrudescense in September. It has been shown that the cobra also exhibits two peaks of testosterone production at these same times. In *Natrix maura* males show peaks of thyroid stimulation in spring and autumn, and females have peaks that correspond to periods of mating and ovulation. The thyroid is not in a resting stage at the beginning of hibernation but colloid accumulation becomes marked during the winter.

Increased activity of the thyroid during sexual activity is found in both males and females of *Clemmys caspica leprosa*, and administration of either male or female hormones to immature specimens is reported to cause activation of the thyroid as well as of the sexual system.

It is difficult to assess the degree to which seasonal changes in thyroid histology or physiology can be taken to demonstrate a direct association between the thyroid secretions and reproductive events. Miller (1959) has pointed out that the period of breeding corresponds with a time of greatest bodily activity (feeding, fighting, etc.) in reptiles, and the thyroid secretions may increase in association with this rather than with any specific reproductive event. Unfortunately experimental studies that would serve to elucidate this matter are still scanty. Mellish and Meyer (1937) found that thyroxin injection causes ovarian atrophy in the horned lizard (*Phrynosoma*), and Evans and Hegre (1938) reported evidence for increased thyroid activity after theelin administration in *Anolis*.

There is no clear evidence for increased thyroid function during pregnancy in live-bearing reptiles. No appreciable changes in thyroid histology occur during gestation in *Vipera*. Eggert (1935) reported that thyroid activity is high throughout gestation in *Lacerta* but attributed this to the tendency of the pregnant females to bask in the sunlight. In *Xantusia* thyroid activity declines after the second or third week of pregnancy and remains relatively low thereafter.

Thyroid and Metabolism

Whether the thyroid plays a role in the control of metabolic rate in reptiles, or indeed in any poikilotherm, has long been open to question. Several early investigators maintained, on the basis of indirect evidence, that some of the features that seem related to changes in thyroid function in reptiles could be interpreted as accompaniments of thyroid-caused alterations in metabolic rate. However, until recently, attempts to demonstrate direct effects of T_4 administration or thyroidectomy upon oxidative metabolism were unsuccessful. The explanation for these failures now seems clear. It has been shown that in *Anolis carolinensis*, although thyroidectomy or administration of T_4 or TSH cause no change in oxygen consumption in animals maintained at 21° to 24°C, all of these treatments have striking effects in lizards kept at 30°C. This suggests that the tissues may be more responsive to thyroid hormone at 30°C (which is near the optimal temperature for *A. carolinensis*) than they are at lower temperatures. This seems to be the case for, in *Eumeces obsoletus* maintained at 30°C and given daily injections of T_4 for three weeks, determinations of oxygen consumption of individual tissues (liver, brain, heart, lung) show significantly higher levels as compared with those for controls. Only skeletal muscle failed to show a response, possibly because the technique used resulted in a low survival of the muscle tissue. The relation between environmental temperature and the metabolic response to thyroid hormone has been convincingly demonstrated in *Lacerta muralis* by maintaining groups of T_4-injected and control animals at 30°C for two weeks, by which time a striking difference in oxygen consumption was demonstrable, and then reducing the temperature to 20° while still continuing the T_4 injections. Within nine days the oxygen consumption in the two groups was at the same level. Returning the lizards to 30°C resulted in reestablishment of a marked difference within a week. In *Eumeces fasciatus* similar results are obtained with T_4 administration, and thyroidectomy, causes a corresponding lowering of oxygen consumption at 30°C but not at 20°C. *Eumeces* given daily T_4 injections and subjected to 10 hours at 33°C and 14 hours at 20° each day show a rise in oxygen consumption after one or two weeks.

Thyroid and Ecdysis

A relation between thyroid function and moulting of the outer layer of the skin is well established for *Lacerta*. This was first indicated by the work of Drzewicki (1926, 1927, 1929) which showed that thyroidectomy causes cessation of moulting. In thyroidless animals the horny layer is formed continuously and the differentiation of a border sheet (*stratum terminativum*) fails to occur. Eggert's (1933, 1936a) extensive investigation of this phenomenon in *Lacerta agilis*, *L. muralis*, and *L. vivipara* showed that the thyroid is relatively inactive during the phase of rapid formation and cornification of new epidermis. During the brief period of actual moulting, however, the height of the thyroid epithelium increases, chromophobe droplets become numerous in the colloid, and colloid release occurs. This activity continues for several days after ecdysis but gradually decreases to reach a low point again during the period preceding the next moult. Eggert confirmed the fact that thyroidectomy causes cessation of ecdysis and also found that implantation of thyroid tissue into the musculature of thyroidectomized lizards enables them to carry out several successful moults. Other studies on *Lacerta* confirm these findings. Recent histological studies on the moulting cycle and its relation to thyroid activity in the lizard *Gekko gecko* also agree with this general picture.

Later investigations concerning other lizards gave less clear-cut results. Noble and Bradley (1933) reported that neither thyroidectomy nor hypophysectomy causes cessation of moulting in *Hemidactylus brookii*, although either operation does result in a lengthening of the period between moults, an effect which can be eliminated by thyroxin administration. Ratzersdorfer et al. (1949) found that thiourea administration to *Anolis carolinensis* has no detectable effect on ecdysis, even though the thyroid shows definite histological indications that effective inhibition of thyroid function has been achieved. On the other hand, in Wilhoft's (1958) experiments with *Sceloporus occidentalis* kept continuously at: 34.0° to 35.0°C it was observed that shedding, in the form of gradual flaking off of the skin, occurred more frequently in the experimental animals than in controls and was correlated with the time of greatest increase in height of the thyroid epithelium. Recent findings by Chiu et al. (1967) show that thyroidectomy decreases the frequency of moulting in the Tokay (*Gekko gecko*) whereas thyroxine injection into unoperated animals increases the frequency. The effects of thyroidectomy may be eliminated by administration of exogenous thyroxine. Hypophysectomy is similarly effective in decreasing the frequency of ecdysis, and this effect can be eliminated by injection of either TSH or prolactin. In this lizard it appears that the thyroid secretion affects only the resting phase of the sloughing cycle and has no effect upon the keratinization phase. Thus no hyperkeratosis occurs after thyroidectomy.

For snakes the evidence thus far indicates that the thyroid influence on ecdysis is the reverse of that in lizards. Schaefer (1933) found that either hypophysectomy or thyroidectomy results in a significant increase in molting activity in *Thamnophis*, while thyroid administration, causes a cessation of shedding. Results leading to a similar conclusion were reported by Krockert (1941) for *Python*, by Halberkann (1953, 1954a, and b) and Goslar (1958a and b, 1964) for *Natrix*, by Maderson et al. (1969) for *Ptyas*, and by Chiu and Lynn (1970) for *Chionactis*. Saint Girons and Duguy (1962a) note some indication of brief periods of increased thyroid activity corresponding to moulting in *Vipera aspis* but do not feel that their studies are decisive on this point.

Ecdysis in squamates is also reported to be influenced by various other substances, notably thymus extract, growth hormone, ACTH, various sex hormones, and prolactin. Whether these act directly upon the sloughing cycle or through some influence on the thyroid gland remains to be investigated.

Thyroid and Growth and Differentiation

There are no reports of comprehensive experiments concerning the effect of thyroidectomy, of thyroid inhibition, or of T_4 or TSH administration upon growth rate in young reptiles. Drzewicki (1929) gave some incidental observations indicating that thyroidectomy causes growth inhibition in the young of *Lacerta agilis*, and Giusti (1931), on the basis of a single specimen of *Clemmys caspica leprosa* thyroidectomized at the age of three months, reported a marked retardation in growth rate in this animal as compared with a control of the same initial age and size. Wilhoft (1958) found that juveniles of *Sceloporus occidentalis* show a higher thyroid epithelium throughout most of the summer than do adults which may indicate a relation to their high growth rate. Krockert (1941), however, observed that a young python fed regularly on pig thyroid gland showed an inhibition in growth rate as compared with a control fed on pig pineal or one fed on rodents.

Inhibition of thyroid function during embryonic stages has been studied in *Chelydra serpentina*. Developing eggs were treated with thiourea solutions either by injection into the albumen or by raising them on cotton moistened with the solutions. Thyroids of these embryos showed the typical response to goitrogens indicating that the concentrations used were effective in inhibiting production of thyroid hormone. The embryos had a decreased growth rate, abnormalities of the carapace were common, hatching was delayed, and normal retraction of the yolk sac into the body did not occur.

Thyroidectomy of pregnant females of *Lacerta* six to eight weeks before the end of term often results in premature discharge of the eggs which, however, contain normal embryos. If the eggs are retained, some embryos degenerate and others survive but are unable to hatch. Thyroidectomy at later stages in pregnancy is followed by more normal development and hatching.

Thyroid and Other Endocrine Organs

In addition to the direct control of thyroid function by pituitary-produced TSH and the relations between the thyroid and gonadal activity which have been discussed above, there are also correlations between thyroid function and various other endocrine mechanisms.

Early studies showed that histological changes occur in the parathyroid glands of lizards after thyroidectomy (*Uromastyx acanthinurus*, *Lacerta agilis*). These changes involve disappearance of some cellular elements and formation of colloid-tilled epithelial vesicles that are not normally present. They are interpreted as indications of hyperfunction. It is noteworthy, however, that this parathyroid response was also seen in lizards in which the thyroid had been removed and then reimplanted.

The ultimobranchial bodies (postbranchial bodies) were reported to undergo hypertrophy in thyroidectomized lizards failed to confirm this response, but found instead that the ultimobranchial bodies show seasonal variations in secretory activity corresponding to those that occur in the thyroid. Moreover they respond to TSH administration in the same way as the thyroid. Long-continued treatment with TSH causes the formation of cysts. Eggert concluded that the ultimobranchial bodies of lizards are directly affected by TSH, but noted that hypophysectomy does not produce any detectable change in these structures. The ultimobranchial bodies of *Lacerta agilis* usually begin to undergo degenerative alterations during the second or third summer. Removal of ultimobranchial bodies causes no observable effect upon the behaviour or physiology of the lizard or upon, the histology of its thyroid.

Removal of the parietal eye in *Sceloporus occidentalis* and *Uma ornata* results in thyroid stimulation. Parietalectomized animals maintained in the laboratory at the mean temperature of their normal habitat show heightened thyroid epithelium and reduction of stored colloid within several months after operation. Lizards allowed to live free in the field after parietalectomy give histological evidence of elevated thyroid activity when collected in winter or spring but not when collected during the summer. The operation does not prevent the thyroid from going through an annual cycle of activity. It is suggested that the reptilian parietal eye may produce a hormone which exerts an inhibiting effect upon the bodily activity of the animal, probably via the pineal, pituitary, thyroid, and other endocrine glands. Studies of the histology and ultrastructure of the parietal eye seem to support the idea that it has a secretory function.

Destruction of the epiphyseal region in the turtle, *Clemmys caspica leprosa*, by electrocautery results in exophthalmia which can be relieved by the administration of lyophilized mammalian epiphysis. The operated animals also show testicular inactivity and thyroid stimulation. There is a marked increase in neurosecretory material in the paraventricular nucleus of tortoises within five months after destruction of the epiphysis. The ultrastructure of the epiphysis in turtles (*Pseudemys*) and snakes (*Natrix*) is consistent with secretory function.

Miscellaneous Effects of Thyroid Hormone

Thyroidectomy is reported to result in formation of fluid-filled cysts in the thymus, and injection of mammalian thymus extract is said to cause thyroid inactivity accompanied by increased frequency of ecdysis in a snake.

Thyroidectomized *Lacerta* show a gradual decrease in physical activity and appetite and they die within three to eight months with indications of an anemia due to decreased blood-forming activity in the bone marrow. Thyroxin administration prevents the appearance of these symptoms. A relation between hematopoiesis and thyroid function is also indicated by the work of Charipper and Davis (1932) on *Pseudemys* and that of M. C. Saint Girons (1961) on *Vipera*. Maher and Levedahl (1959) concluded that the high mortality in their experiments with thyroxine-treated *Anolis* was related to involvement of the nervous system since the animals were subject to uncontrolled muscle contractions after tactile stimulation.

General pugnacity and territorial behaviour are increased by thyroxin or TSH in *Anolis* and thyroid feeding increases irritability in *Python*.

Conclusion

During the past decade there has been a notable increase in researches concerning the thyroid of reptiles, and these have resulted in new insights into both the morphology and physiology of the gland in this group. In addition, these studies have served to indicate a number of promising areas for future investigation.

Newly developed histological techniques should be more fully applied to study of both the thyroid and pituitary in an effort to further elucidate questions concerning the secretory process, the precise identification of thyrotrophs, and the feedback mechanism involved in pituitary control. Autoradiographic determination of I^{131} localization in the cells of the thyroid epithelium and in the colloid under diverse conditions are not yet available for any reptile. Such studies would be of particular interest for the macrothyreocytes of snakes. Investigation of the ultrastructure of reptilian thyroid cells has made but a bare beginning, but there are indications that use of the electron microscope on such material will be most rewarding. Again, examination of the macrothyreocyte as a possible site of intracellular colloidogenesis seems especially needed. No work has been done on reptiles on the cellular biochemistry of the thyroid as revealed by study of enzyme systems, nor is there any information for this group concerning blood transport of thyroid hormones or their utilization in the peripheral tissues.

Although the origin and differentiation of the thyroid have been investigated for a number of reptiles, much remains to be done in the field of development. None of the lizards in which the thyroid is completely paired have been studied embryologically. It would be interesting to know when and how the originally single thyroid Anlage separates into two parts and to attempt to ascertain whether there is any mechanical explanation for this separation, as seems to be the case in anurans. Thus far there has been no approach to experimental study of reptilian embryology by microsurgical techniques. This is doubtless due to the lack of readily available embryonic material, but it would seem that the familiar methods that are so widely used in chick embryology could be adapted for reptilian eggs. Repetition of much of the classical work on avian embryology dealing with embryonic induction would then be possible and would be of special interest in reptiles, the only poikilothermous amniotes. With respect to the thyroid, a study of the differentiative capacity of the thyroid Anlage in chorioallantoic grafts immediately suggests itself. A related field of experimentation is the culture of embryonic thyroid tissue *in vitro*, another promising method for investigation of secretory function which remains to be exploited in these animals. The earliest beginning of iodide concentration by the cells of the embryonic thyroid, as indicated by uptake of I^{131}, has been studied for turtles only, and much more needs to be done in this area.

The relation between I^{131} metabolism and environmental temperature is known for several lizards, but more work is needed comparing this feature in reptiles living in regions in which optimal temperatures are widely different. The possible role of the thyroid in high temperature tolerance may be susceptible of elucidation by such studies, and this may be closely related to the mechanism of adaptation to special habitats.

Maher's (1961) recent studies on the metabolic effects of the thyroid hormone in lizards kept at high temperature have effectively disposed of the old question of whether the thyroid ever has any effect on metabolic rate in any poikilotherm. Such work must now be extended to other reptilian groups and to a wider variety of experimental treatments.

The extensive investigation by *Eggert* (1935, 1936a and c) concerning the role of the thyroid in ecdysis in *Lacerta* remains the classic in this area, and his findings are unequivocal. However, the conflicting results obtained in the few available studies on other lizards and the evidence for an opposite effect in snakes should lead to new comprehensive work on this phenomenon in other squamates.

The effect of the thyroid secretion upon bodily growth and differentiation is another subject that has been largely neglected. Much more work is needed directed towards accurate determination of effects of T_4 administration, thyroidectomy, and other experimental treatments in young animals and during embryonic stages. Such studies must concern, not only general body growth but also growth and differentiation of various organ systems like the nervous system, skeletal system, and reproductive system which are known to be influenced by the thyroid in other animals.

The work of Eggert (1936b) on the thyroid-like responses of the ultimobranchial bodies in *Lacerta* is an example of a particularly interesting early observation that deserves further attention. Effects of the thyroid upon hematopoiesis in reptiles have been indicated by several studies, but many other physiological relationships that are well demonstrated in mammals have not yet been explored in any reptile. These include renal function, secretory and muscular activity in the gastrointestinal tract, calcium metabolism, functioning of skeletal muscles, and wound healing.

Finally, the relation of thyroid function to behavioural patterns in reptiles is a field that is undoubtedly destined to develop rapidly. The studies of cyclic changes in the thyroid in connection with the reproductive cycle indicate that thyroid activity may play a part in sexual behaviour as well as in gonadal maturation and ovulation. The results of parietalectomy in lizards suggest a complex role of the parietal eye, the thyroid, and other endocrine organs in the behavioural patterns involved in seeking optimal conditions of light and temperature. There is currently much active interest in both of these aspects of reptilian behaviour, and the part played by the thyroid must certainly receive further evaluation.

4

Parathyroid Gland

The parathyroid glands of mammals have been studied intensively because of their important role in calcium and phosphate homeostasis. Much less is known of the parathyroids of members of other vertebrate classes, and reptiles are among the least well studied in this respect. This paper is a review of our present knowledge concerning the location, structure, and function of the parathyroid glands of reptiles.

The structure of the reptilian parathyroid is similar to that found in mammals, birds, and urodeles, but is quite different from the whorl-like arrangement of cells in the parathyroid glands of anurans. Reptiles have two or four parathyroid glands, depending upon the group, species, or age of the animal studied. There is still confusion regarding the number of parathyroids in some species, as various epithelial glands described by early authors have sometimes been unreliably interpreted to be parathyroid tissue. The German term "Epithelkorper" has contributed to the confusion; in the early German literature the word referred to a variety of small nodules. Even more confusion has been caused over the years by calling the parathyroids or other epithelial structures "Carotiskorperchen" or "carotid bodies".

Studies of embryological development of the parathyroids of reptiles generally indicate that the rostral and caudal pairs of parathyroids are derived from pharyngeal pouches three and four respectively. The parathyroid glands are not associated with the thyroid glands, but are located nearer the thymuses or ultimobranchial bodies because, unlike the mammalian parathyroids, they do not migrate, but tend to retain their original relationship with the third and fourth aortic arches. In those animals having four parathyroids, the anterior (or rostral) pair is generally located near the branching of the carotid artery, except where, as in the Testudines, the definitive carotid bifurcation is developmentally a secondary one; the posterior (or caudal) pair is located near the aorta. When only a single pair of glands persists in adults, it is that derived from pouch three.

The parathyroid glands of reptiles are small, measuring about 0.5 to 1.0 mm in diameter, and are easily overlooked in dissections. Their structure is essentially similar in the various groups of reptiles, and parathyroid tissue is easily recognized in histological preparations. The parathyroids are surrounded by a substantial capsule of connective tissue and have a cellular, cord-like parenchymal structure with connective tissue and capillaries or sinusoids present between the cords. Most reports of their histology indicate that the glands contain a single type of cell, presumably homologous with the chief cell of mammals. Oxyphil cells appear to be absent.

A unique feature of the glands in several groups of reptiles is the presence of "follicles" consisting of cells surrounding a central lumen. In many cases these follicles contain a secretion which stains positively for carbohydrate content. There is speculation that this may represent stored parathyroid hormone, but experimental evidence is not yet available. Follicles occasionally occur in mammalian parathyroid tissue, but generally only in older animals, in which they have been considered an effect of age. However, follicles have been noted in the parathyroids of young turtles and, therefore, cannot be dismissed as an aging phenomenon in reptiles.

Little work has been done to determine the function of the parathyroid glands of reptiles. However, the few studies performed indicate at least some similarities of reptilian and mammalian parathyroid function in the regulation of the concentrations of calcium and phosphate in body fluids.

Testudines

Although the literature before 1911 contains several conflicting reports, turtles possess two pairs of parathyroid glands, which develop from pharyngeal pouches three and four. The anterior pair of glands is embedded within the tissue of the thymus and is difficult to detect except upon histological investigation. The posterior pair of glands is more obvious, lying in connective tissue near the arch of the aorta and in close association with the tissues of the left ultimobranchial body. Some early papers refer only to the more obvious posterior pair of glands; the anterior parathyroids, located within the thymus, were

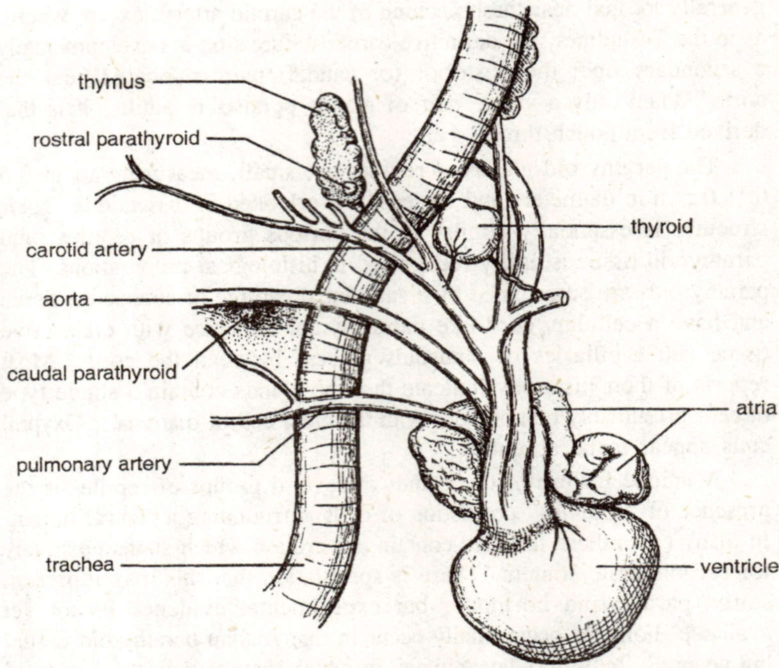

Fig. 4.1. Location of the right parathyroid glands in a freshwater turtle, Emys orbicularis.

apparently overlooked. In 1911, Aime commented that some earlier workers who performed parathyroidectomies in turtles failed to remove both pairs of parathyroid glands.

The parathyroid glands of turtles measure about 1 mm in diameter in adults and about half that size in hatchlings. The histology of the parathyroids varies somewhat from one species of freshwater turtle to another. While the parathyroid parenchyma of all species studied consists of cellular cords, a conspicuous follicular arrangement of cells is characteristic of the slider, *Pseudemys scripta*, and the false map turtle, *Graptemys pseudogeographica*. However, fewer follicles are found in parathyroid tissue of the painted turtle, *Chrysemys picta*. Some of the follicles contain a substance which stains with periodic acid Schiff (PAS) but is not glycogen.

The cells of the parathyroid glands of these three species of turtles, as seen with the light microscope, appear to be of a single type. They contain round or oval nuclei and a finely granular cytoplasm which stains faintly with PAS and eosin. Use of a technique that stains mammalian mitochondria did not demonstrate cells rich in mitochondria which might correspond to the oxyphil cells of mammals. In silverstained preparations, nerves were seen entering the thymus and ultimobranchial bodies of turtles, but no nerves were seen in connection with the parathyroid tissue. Accessory parathyroid glands were not found in histological sections of the region near the heart and major blood vessels in the turtles studied.

We have performed preliminary tests of the histochemistry and ultrastructure of the parathyroid glands of the freshwater turtles *Chrysemys picta* and *Pseudemys scripta*. Tissues for histochemical study were fixed in Bouin's fluid or formalin and embedded in paraffin or plastic, while tissues for electron microscopy were fixed in osmic acid (2%) or in glutaraldehyde and post-fixed in osmium and embedded in plastic. The following histochemical tests were performed for polysaccharides: alcian blue for acid mucopolysaccharides, toluidine blue for metachromasia and mucin, and periodic acid-Schiff. All but alcian blue were positive, indicating that a mucoid or mucopolysaccharide component may be present in the cytoplasm and in the contents of the follicular lumen. The mercury bromphenol blue test indicated a proteinaceous component in both cytoplasm and follicular lumen. The most striking histochemical finding was the demonstration, by the Sudan Black B method of McGee-Russell and Smale (1963), of large lipid droplets in the cytoplasm surrounding the nucleus. This test gives a reversed or negative effect, in which the epon plastic becomes stained while the dense elements of the tissue (nuclei) do not; thus the nuclei appear white, the cytoplasm grey, and the lipid droplets black. Ultrastructural studies verify the presence of large lipid droplets in the cytoplasm, as can be seen in the low magnification of a cell cord.

The cytoplasm of some of the parathyroid cells is electron lucent, while that of others is electron dense. It is not yet known if these represent different types of cells or different secretory stages of a single type. The light and dark cells are especially well seen in glutaraldehyde-fixed material.

The cisternae of the endoplasmic reticulum of both the light and dark cells sometimes contain an electron dense material. Under high magnification, this material appears as parallel electron-dense bands which may represent crystalline material embedded within a less dense matrix; its structure is similar to that described for the endoplasmic reticulum of salamander liver. Another feature of the parathyroid cells of both species of turtles is membrane bound, electron dense granules about 3000-4000 Å in diameter. The granules possess a substructure near the centre which closely resembles that seen within the endoplasmic reticulum. Further studies are in progress to determine the nature and function of the various cytoplasmic inclusions.

There have been only two studies of parathyroid function in turtles. The former performed parathyroidectomies on the "African tortoise" by removing a single pair of glands in the region of the aortic arch and pulmonary arteries. The result was paralysis and death of the animals. Removal of one gland was without effect. Since these authors did not measure blood calcium values, it is not certain that the results were due to parathyroid loss or innumerable other factors which might affect nerve function. It is impossible to repeat the experiment or to know if complete parathyroidectomies were performed, as the species of turtle used is not certain.

I was unable to induce tetany, paralysis, or significant alterations in total concentrations of calcium or inorganic phosphate in the serum of the freshwater turtles *Chrysemys picta* or *Pseudemys scripta* after removal of both pairs of parathyroid glands. The concentration of calcium in the serum of parathyroidectomized and normal animals remained at about 2.5 mM/L, while that of inorganic phosphate was approximately 1.0 mM/L in both groups. There were considerable fluctuations in these values from day to day in an individual, whether parathyroidectomized or sham-operated.

The one parameter which was altered by parathyroidectomy of the turtles was urinary excretion of phosphate which decreased in parathyroidectomized animals compared to its high levels in sham-operated controls. A similar hyperphosphaturia in sham-operated rats has been reported by Beutner and Munson (1960). Correlated with this was a significant rise in urinary excretion of phosphate in parathyroidectomized and normal turtles which received 100 IU or more of mammalian parathyroid extract. After a three-day latent period, they excreted significantly higher amounts of phosphate, up to six times the normal amount by six days after the injection. The same animals had 20% higher concentration of calcium in their serum, again after a three-day latent period, but in this case the differences were not statistically significant ($P=0.1$). There was no significant alteration of phosphate concentration in the serum or in the urinary level of calcium.

Administration of parathyroid extract did not affect the concentrations of calcium in fibulae as measured by wet-ash analysis. However, femurs of hatchling false-map turtles, *Graptemys pseudogeographica*, injected with parathyroid extract contained an average of one third more osteoclasts than control animals.

The reasons for the unresponsiveness of turtles to parathyroidectomy remain to be elucidated. It has been suggested that these animals might be able to obtain sufficient amounts of calcium and phosphate by means of a physicochemical exchange between the large bony surface of the shell and the body fluids, or that they may have active transport systems for calcium in their pharyngeal or cloacal tissues.

CROCODILIA

The best description of parathyroid location and embryological derivation in the Crocodilia is that of Van Bemmelen (1886 and 1888). He described a single pair of parathyroids derived from pouch three in specimens of *Alligator mississippiensis* and *Crocodylus porosus*.

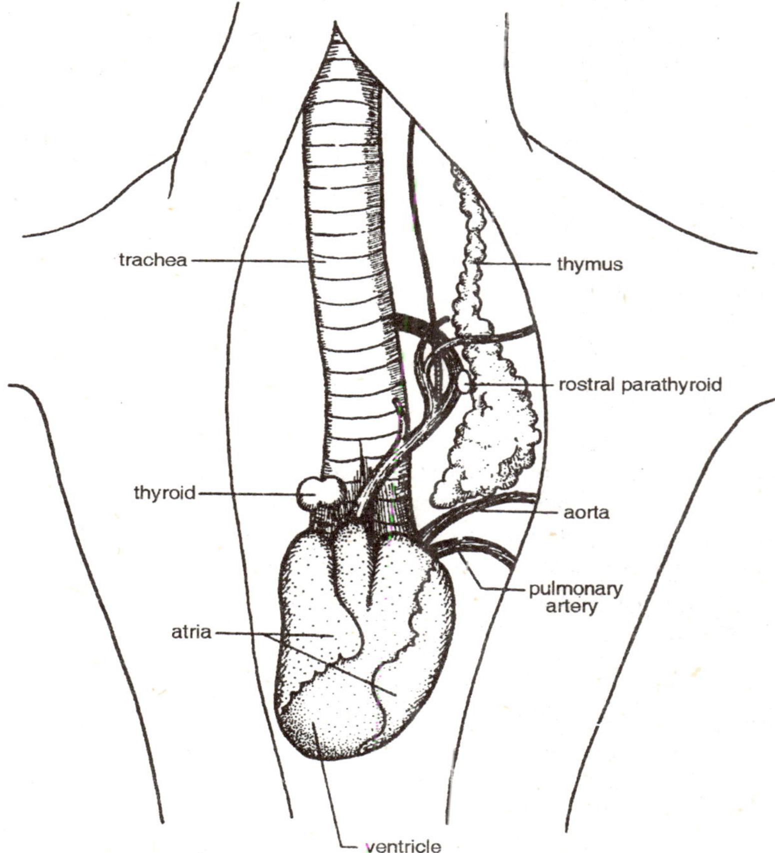

Fig. 4.2. Location of the left parathyroid gland in Crocodylus porosus.

Hammar (1937) reported that two pairs of parathyroid glands were present in young crocodiles, but that the pair from pouch four disappeared in adults. The parathyroids of the Crocodilia are located just anterior to the heart, where the common carotid artery branches from the right innominate artery. They are not intimately connected with the blood vessels.

As there are no previously published accounts of the parathyroid histology of Crocodilia, four hatchling caimans (*Caiman crocodilus*; snout-vent length 13 cm) were examined. In all four animals, the parathyroid tissue on each side was composed of two glands surrounded by a single connective tissue capsule. The size of the smaller gland ranged from about 0.25 to 0.5 mm in diameter, while the larger gland measured approximately 1 mm. It is probable that the double parathyroids represent the glands from pouches three and four described by Hammar (1937), and that later one pair (presumably the smaller) would have disappeared or fused with the other.

The parathyroid glands of the caimans lie near the branching of the right innominate artery and may sometimes be embedded within the thymic tissue. The glands are composed of cellular cords with connective tissue and blood vessels running between them. The cells have centrally located nuclei, and the cytoplasm stains with PAS and eosin. The few follicles do not form a conspicuous feature of the glands.

Lepidosauria

Rhynchocephalia

The parathyroids of the rhynchocephalian, *Sphenodon punctatus*, have been the object of only a few studies. Van Bemmelen (1887 and 1888) described a pair of "Carotiskorperchen" and a slightly more posterior pair of Aortakorperchen in adult *Sphenodon*. Later investigators have interpreted these as two pairs of parathyroid glands. Described as small, round bodies of epithelial structure, the Carotiskorperchen are located at the bifurcation of the carotid artery into internal and external carotids while the Aortakorperchen lie close to the posterior wall of the aortic arch. From his studies of their development in *Lacerta*, Van Bemmelen (1887) decided that the glands in *Sphenodon* originate from pharyngeal pouches three and four respectively.

Only Gabe and Saint Girons (1964) have recently investigated parathyroid histology in *Sphenodon*. They confirmed the presence of four parathyroid glands in their two specimens. The parathyroids are

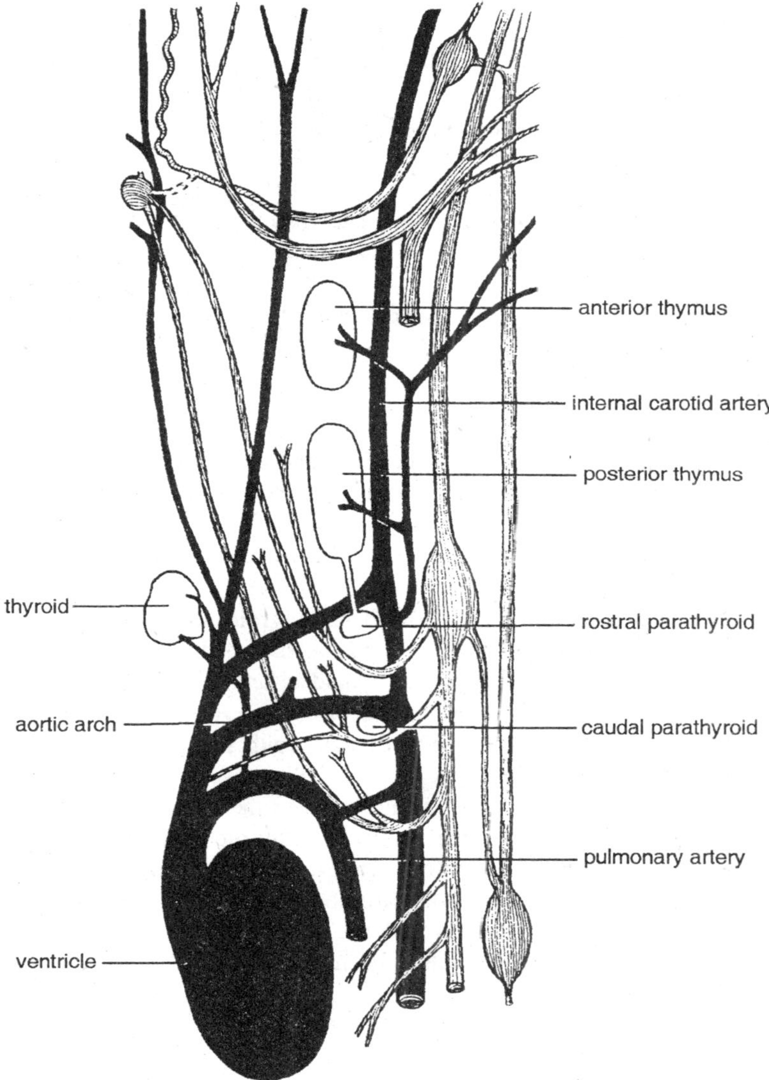

Fig. 4.3. Location of the left parathyroid glands in Sphenodon punctatus.

structurally similar to those of other reptiles, being composed of cellular cords invested with connective tissue strands and capillaries. The parathyroid tissue contains only a single type of cell which measures about 12-15 μ in diameter, contains a central nucleus and has cytoplasm containing PAS-positive granules which were identified as glycogen by the malt diastase digestion technique. The parathyroid cells are not arranged in follicles.

The fact that *Sphenodon* possesses four parathyroid glands was, at first, interpreted as evidence of its primitive nature and was even regarded as a specific character, as most lizards have only a single pair. Subsequent investigations have shown that many other reptiles, including snakes, turtles and some lizards, also have four parathyroid glands.

There have been no physiological investigations of the parathyroid glands of *Sphenodon*.

Squamata

Ophidia

Early work on the number, location, and embryological origin of the parathyroid glands in snakes was done by Van Bemmelen (1886 and 1888), Verdun (1898), Saint-Remy and Prenant (1904), and Harrison and Denning (1929). There is still confusion regarding their embryological derivation.

The most complete description of parathyroid numbers, location, and histology has been provided by Herdson. In a study of seven species of snakes of the families Colubridae, Elapidae, and Viperidae, he found two pairs of parathyroids in every case, with no evidence of accessory parathyroid tissue. The rostral pair of parathyroids lies at the bifurcation of the carotid arteries near the angle of the jaw, while the caudal pair lies a considerable distance away, between the lobes of the thymus just anterior to the heart. The caudal parathyroid glands of *Vipera berus*, located close to the thymus glands, are diagrammed by Boyd (1942). The parathyroids of snakes are not in intimate connection with the walls of the blood vessels and are not innervated. The glands possess connective tissue capsules and measure about 0.5 to 1.0 mm in diameter. Ophidian parathyroids are formed by cellular cords, with connective tissue and capillaries lying between the cords. The nuclei are large and the cytoplasm scanty.

Four species of snakes, *Thamnophis sirtalis*, *Coluber constrictor*, *Natrix sipedon*, and *Lampropeltis doliata*, have been studied in my laboratory. Their parathyroid glands are about 0.5 to 1.0 mm in diameter and have the typical parenchymal structure with cellular cords. The cytoplasm stains faintly with PAS and contains PAS-positive granules, histochemical tests performed on the parathyroids of *N. sipedon*

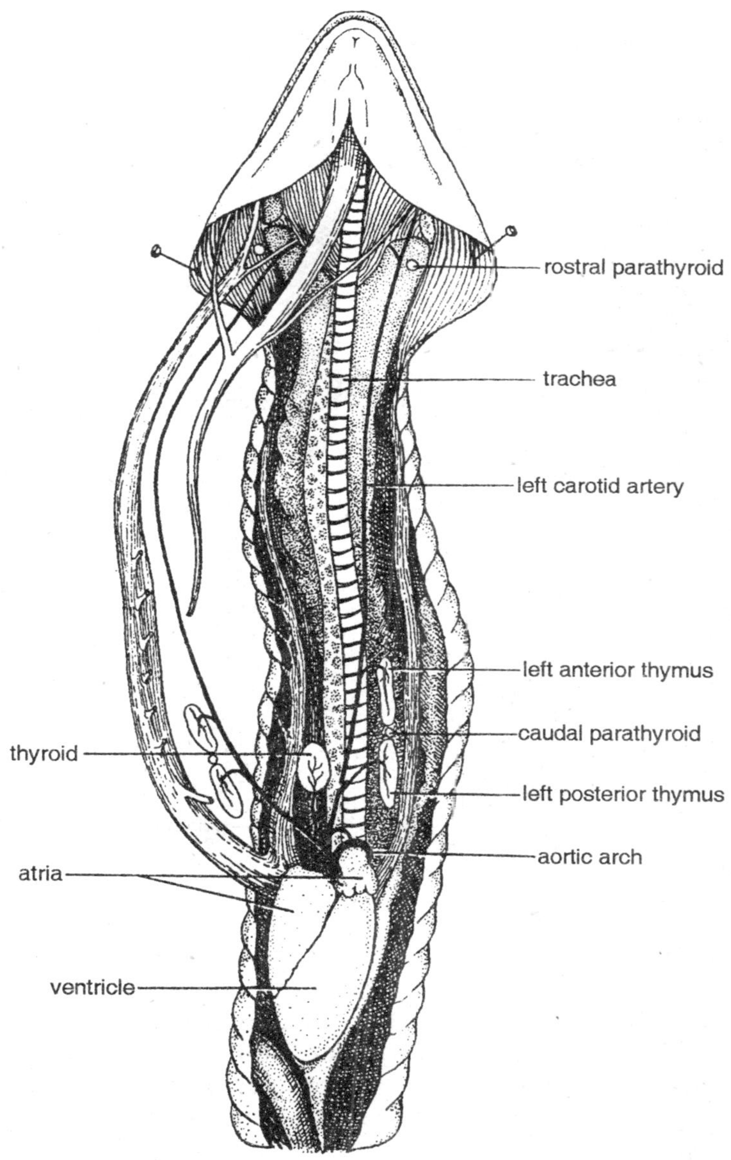

Fig. 4.4. Location of the parathyroid glands in the snake Natrix natrix.

demonstrated the presence of protein in the cytoplasm but no evidence of lipids. Follicles are sometimes conspicuous in the parathyroid glands

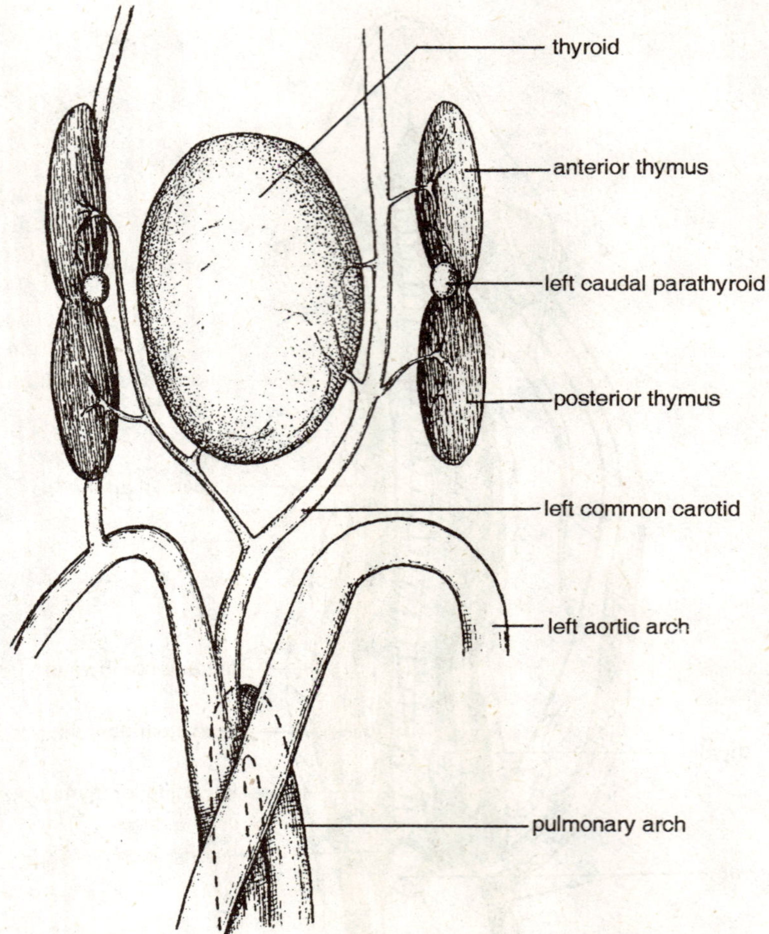

Fig. 4.5. Detail of the area of the thymus glands in the snake Thamnophis, showing the location of the caudal pair of parathyroid glands.

of *Thamnophis sirtalis*. The glands appear to be composed of a single type of cell, as noted by Herdson (1956). Their locations agreed with those reported by Herdson in all cases but one: in a young *Natrix* the caudal parathyroids were not conspicuous in gross dissection. However, when the thymuses were removed and sectioned, a parathyroid gland was found closely adhering to the surface of each thymus. In this individual, the thymuses which were about 1 cm long, several times the size of those in older individuals, appear to have almost enveloped the caudal parathyroids. In an earlier paper, Thompson (1910) noted a parathyroid gland embedded in the head of the thymus in *Natrix sipedon*. Her finding that the parathyroid glands of this species lack follicles and lumina is confirmed by my observations.

We have recently parathyroidectomized the snake *Thamnophis sirtalis* in my laboratory. Blood samples were collected from the tail, ten days after the operation. Their values for concentration of total calcium were 30% lower than those of the sham-operated controls. The parathyroidectomized animal with the lowest blood calcium (>50% below normal) went into tetanic convulsions six days later. A snake from which three parathyroid glands had been removed went into tetany 51 days later, but none of the animals died as a result of the operation. The evidence suggests that these glands are important to the calcium metabolism of snakes.

Sauria

The parathyroids of lizards are better known than those of any other group of reptiles. Most adult lizards possess a single pair of parathyroid glands, the rostral pair derived from pharyngeal pouch three; a caudal pair, derived from pharyngeal pouch four, generally develops during the early embryonic stages and subsequently disappears. Recently, however, several species of lizards have been found to possess two pairs of parathyroid glands, and possibly this condition will be found in other species as well.

In those lizards possessing a single pair of parathyroids, the glands are located near the bifurcation of the carotid arch in the cervical region. The glands are generally closely associated with the blood vessels and may be embedded in the adventitia, thus making parathyroid removal rather difficult without damage to the blood vessels. When a second pair of parathyroids is also present, it is about half the size of the rostral pair (approximately 0.5 mm diameter) and located just posterior to them near the arch of the aorta on either side of the neck.

The histological appearance of the parathyroid glands has been described for many species of lizards and resembles that of other reptiles. The glands are composed of cellular cords, with capillary or sinusoidal networks between the cords, and are surrounded by a connective tissue capsule. The parathyroid cells have round or oval nuclei that are rather large relative to the amount of cytoplasm. The cytoplasm contains only faint granulation and stains weakly with PAS. A characteristic feature of some lacertilian parathyroids is the presence

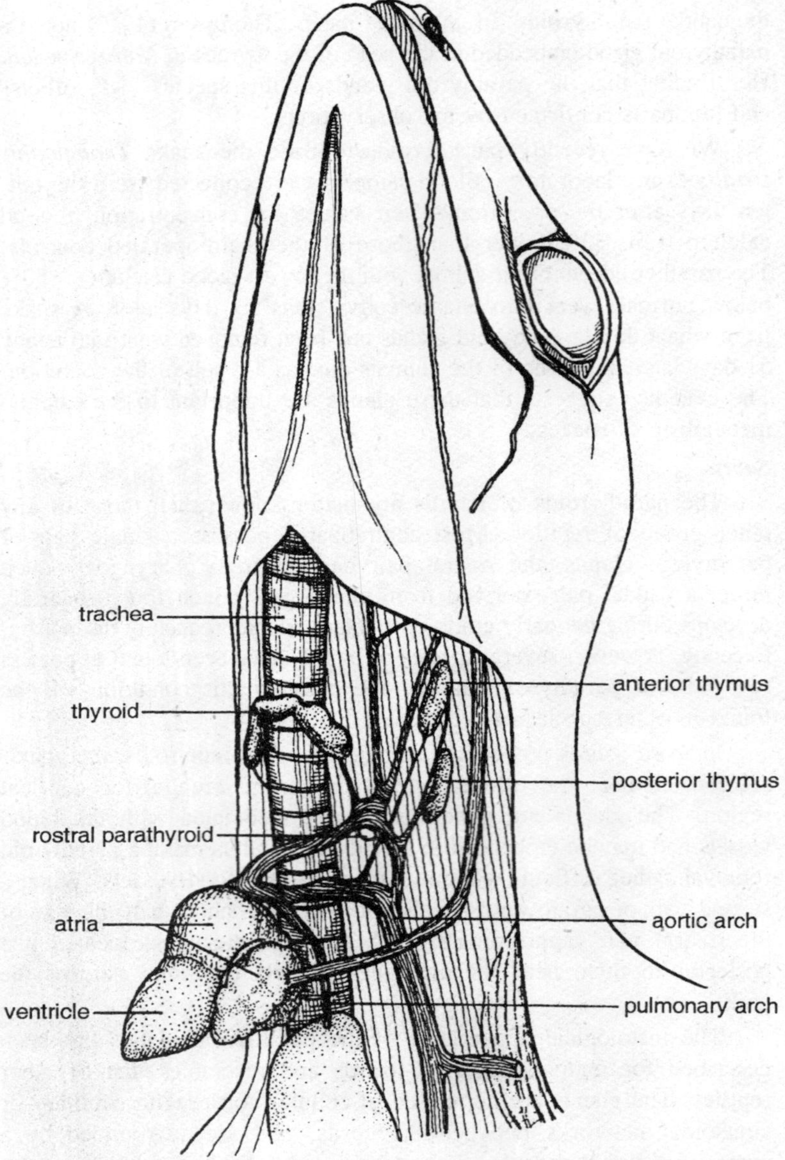

Fig. 4.6. Location of the left parathyroid in the lizard Gekko verticillatus.

of follicles. The follicles often contain material which stains strongly with PAS.

Although light microscopy of the parathyroid suggests that there is only a single type of cell, Rogers (1963), on the basis of the only histochemical investigation of lacertilian parathyroids, has distinguished three types of cells in these glands of the Australian skinks *Tiliqua occipitalis* and *Trachydosaurus rugosus*. The most common type of cell is compared with that of mammals, and given the same name, chief cell. Vacuolated chief cells are compared with the mammalian "water clear" variant and are not considered a separate type of cell. Two rare additional types are the "dark cells", containing smaller nuclei and cytoplasm which can be distinguished histochemically from that of chief cells, and the "epithelial cells", in which both cells and nuclei are fusiform and the cytoplasm is nongranular; the significance of these different cells is not known. Ultrastructural investigations would probably demonstrate whether different types of cells exist in these animals.

Seasonal variation in the histological appearance of the parathyroids has been reported in several species of lizards. Peters (1941) noted some degenerative changes and occasional formation of cysts in *Lacerta viridis* in the winter. About half the parathyroid cells in the African skinks *Chalcides ocellatus* and *Scincus scincus* regressed and disappeared during the winter months and then regenerated during the summer. Such seasonal changes also occur in several species of anurans, and von Brehm (1964) attributed these changes to winter temperatures as they could be induced by subjecting the animals to low temperature out of season. However, Sidky (1965) has attributed the degenerative changes in the parathyroids to failure of his lizards to feed, and has duplicated the effect by starvation of the lizards during the summer months.

Accessory parathyroid glands have occasionally been reported. The relatively frequent occurrence of accessory tissue in a species may reflect the occasional persistence of a second pair of parathyroid glands, as in *Lacerta viridis*, in which Peters (1941) found accessory parathyroid tissue in 6 out of 16 animals. On the other hand, Sidky (1965) found accessory glands in only one of over forty lizards (*Chalcides ocellatus* and *Scincus scincus*) examined and Adams (1939) noted the retention of a caudal parathyroid within the thymus in only one specimen of *Lacerta*.

Parathyroidectomy has been performed on several species of lizards, including *Lacerta viridis*, *Chalcides ocellatus* and *Varanus griseus*, and *Anolis carolinensis*. In all species, removal of the parathyroid glands (one pair in all except *A. carolinensis*, in which there are two pairs) resulted in hyperexcitability and tetanic convulsions, usually after a latent period of several days or more depending upon the environmental temperature. Peters (1941) found that some of his parathyroidectomized animals did not respond as described above, but in most cases histological investigation revealed accessory parathyroid tissue or remnants of the removed pair.

Sidky (1966) was the first to measure the concentration of calcium in the plasma of parathyroidectomized lizards, and found that this falls to approximately half the normal value of 3 mM/L. Blood samples which are taken from parathyroidectomized lizards in tetany tail to clot, indicating the low concentration of calcium ions in the body fluids. Unilateral parathyroidectomy has no effect on the behaviour of the animals or the concentration of calcium in their plasma and administration of calcium gluconate relieves the convulsions and tremors of parathyroidectomized animals.

In *Anolis carolinensis*, removal of three of the four glands usually does not affect the animal's behaviour. Removal of all four glands, however, results in tetanic convulsions and paralysis within one or two days after the operation. The attacks are induced when the animal does any sort of mild exercise, are of about a 30-second duration, and can be reinduced after approximately one hour.

Parathyroidectomized *A. carolinensis* have concentrations of calcium in the serum less than half those of sham-operated controls, and the concentrations of phosphate in their serum are 70% above normal. Parathyroid extract eliminates tetanic seizures of parathyroidectomized animals and causes the concentrations of calcium and phosphate in the serum to return to near-normal levels.

Administration of parathyroid extract induces hypercalcemia, hypercalciuria, and hyperphosphaturia in normal *Dipsosaurus dorsalis* and *Sceloporus grammicus*, but is without effect on the level of phosphate in the serum. I can demonstrate no significant effect of parathyroid extract on concentrations of calcium or phosphate in the serum of normal *Anolis carolinensis*.

In a study of lizard limb regeneration, Umanski and Kudokotzev (1951) demonstrated that the administration of parathyroid extract results in increased numbers of osteoclasts in the bone of *Lacerta agilis*.

Thus lizards yield the best evidence of parathyroid influence on the calcium and phosphate metabolism of reptiles.

5

ADRENAL GLAND

The adrenal gland of reptiles has been known since its description by Perrault (1676) and Morgagni (1763), but its real significance only became recognized in the 19th century. The relatively small number of papers dealing with the reptilian adrenal contrasts markedly with the enormous amount of research that has been devoted to it in eutherian mammals.

The classical publications of the early 19th century give the first detailed descriptions of the microscopic anatomy of the organ and establish its homology with the mammalian adrenal gland. Leydig (1853), Braun (1879), Pettit (1896), Vincent (1896), Minervini (1904) and Poll (1904a, 1904b, 1906) review the numerous histological investigations of the reptilian adrenal, the details of its vascularization, and its interspecific differences. The brief statements these offered seem to have been considered sufficient by herpetologists to establish the topographic position of the reptilian adrenal gland. Those workers interested in histophysiological or physiological approaches seem to have hesitated to study animals difficult to capture, to manipulate, and to maintain live in the laboratory, and have consequently experimented mainly with small mammals. As a result, the number of publications on the adrenal gland of reptiles diminishes sharply at the beginning of the twentieth century.

The lack of information is even more obvious when one attempts to review the physiology of this organ. The two hormones, adrenalin and noradrenalin, produced by the adrenomedullary tissue of the mammals have also been identified in reptiles. The biosynthesis of the corticosteroids in reptiles and the interrelations of their adrenal gland with other endocrine glands are particularly poorly known. Their seasonal and other cyclic changes have been explored in only a very few species.

It is useful to start by stating the nomenclature to be utilized. The term "adrenal gland" must be maintained since it is universally used, but it is misleading. Except among the turtles, the reptilian adrenal lacks any clear anatomical relation to the kidney; in contrast to other vertebrates its relations are much closer to the gonads and to the gonoducts.

The names of the two parts of the organ are even more difficult. Many authors attempt to maintain the nomenclature used for eutherian mammals and designate the cytological and physiological equivalent of their adrenal cortex as "adrenocortical tissue" in reptiles. They also describe a reptilian "medullary" tissue. This approach is clearly inappropriate.

The "medullary" tissue of all reptiles examined occupies a peripheral position, and the "adrenocortical" tissue is centrally placed. The two types of tissue are often mixed, but there are examples in which transverse sections of the organ show a cortex entirely constituted by "medullary" tissue, and a medulla formed of "adrenocortical" tissue only. A comparable anatomical position makes the term "adrenocortical tissue" equally inappropriate in birds and anamniotes. It seems preferable to refer to this tissue as "interrenal", a term created by Balfour (1876) for the cytological and functional equivalent in other vertebrates of the adrenal cortex of mammals.

The term "medullary" is, fortunately, less often applied to the reptilian adrenal; most workers refer to chromaffin tissue, but use of this term also presents problems. The terms "chromaffin" or "phaeochrome" tissue designate all those cells derived from the neural crest and specialized for the production of sympathomimetic catecholamines independent of their localization. However, the homologies of the mammalian adrenomedullary tissue are not with topographical equivalents in other vertebrates. The term "adrenal tissue" is most nearly adequate since it simultaneously defines the nature and the localization of the tissue, and it is hence adopted here.

Embryonic Development

The duality of embryonic origin of the two tissues that form the adrenal gland of reptiles was well established by the classical works published during the end of the 19th and the beginning of the 20th century and agrees with that in all other vertebrates. However, even some of the most recent works contain categorical contradictions concerning the embryology of the reptilian adrenal gland, and the number of species studied in detail is yet small; thus new investigations utilizing the techniques of experimental embryology are still desirable.

Two fundamental questions about the ontogeny of the adrenal gland are: what is the origin of the cells constituting the two tissues, and when do they switch from the embryonic to the functional state ? The first of these seems to have attracted the attention of most of the authors, but our knowledge of both of these must be reviewed.

Origin and Early Ontogeny of the Interrenal Tissue

All authors agree that the reptilian interrenal tissue, like that of all vertebrates, is of mesodermal origin. Yet there are at least two distinct views regarding the establishment of the Anlage of the organ.

Most authors concur that the interrenal tissue is derived from the coelomic epithelium. It differentiates either within the epithelium proper or in the underlying mesenchyme. Proliferation occurs in the prolongation of the genital crest, between this and the root of the mesentery, or on the lateral or medial border of this crest. Some descriptions mention a continuous proliferation; others refer to the production of a series of buds which secondarily fuse.

Weldon (1884) and Hoffmann (1889) presented a very different concept. According to these authors the Anlage of the interrenal tissue derives from the internal wall of the Malpighian corpuscle of the mesonephros; the dorsal part of the resulting blastema develops between the mesonephros and the vena cava and becomes the interrenal tissue, while its ventral portion reaches into the genital crest and forms the testicular net. The connection between the two Anlagen could be noted until a relatively advanced stage of development. Only Raynaud (1962; *Anguis fragilis*) among recent authors reaches conclusions similar, in part, to those of Weldon and Hoffmann.

Miller's (1963) description of the development of the adrenals of *Xantusia vigilis* is typical of the first and more widely held view. According to him, in 2 mm long embryos of *Xantusia vigilis*, cells, destined to form the interrenal tissue in the intermediate mesoderm, proliferate between the future positions of the gonad and the mesonephros. The first identifiable elements seem to derive from mesenchymatic cells located just below the peritoneum lateral to the gonadal Anlage and ventromedial to the mesonephros. Groups of these cells occur on both sides of the midline, lateral to the root of the dorsal mesentery. In 3 mm embryos the interrenal cells have become considerably more numerous, and osmiophilic droplets appear in their cytoplasm. By the 5 mm stage the mass of interrenal tissue has gradually increased and extended posteriorly, and has almost contacted the adrenal element extending from the zone of the paravertebral ganglia. By the 6 mm stage this contact is realized, and the osmiophilic droplets are much more abundant in the interrenal cells. Later developmental stages show a progressive organization of the adrenal cells into a dorsal cap on the surface of the osmiophilic, lipid-rich interrenal tissue which is penetrated by many adrenal cells. Near the time of hatching (22 mm stage) the interrenal tissue and the gonads start to separate, continuing to do so during the first six months of life, as the mesonephros regresses.

Raynaud's (1962) description of the development of *Anguis fragilis* is typical of the second view. In 4.5 to 5.2 mm embryos he notes a proliferation of the external wall of Bowman's capsule of the mesonephric glomerulus lying on the medial side, near the junction of the atrophied nephrostomal duct with the metanephritic vesicula. The proliferation produces a cluster or cells between the medial border of the mesonephros, the peritoneum, and the aorta. By the following stage (embryos from 5 to 6 mm), the cluster has increased in volume, the cells are migrating dorsally, and proliferation of the peritoneum occurs at the contact of its ventral border with the cluster of cells. After this developmental stage it is impossible to determine from cross sections whether the cellular cluster of mesonephritic origin remains autonomous or whether it incorporates peritoneal cells. The interrenal Anlage, however constituted, shows a segmental pattern, and its lateral border adjoins the medial surface of each Malpighian corpuscle of the mesonephros. As development proceeds, the interrenal Anlage shifts progressively away from the peritoneum and reaches a position dorsal to the gonad. Lying between the internal veins of the mesonephros and the medial border of that organ, the Anlage begins to detach from the Malpighian corpuscles in embryos weighing 40 to 60 mg; the separation of the two organs is complete when the embryo reaches 80 to 100 mg.

Both Raynaud (1962) and Miller (1963) comment that only the techniques of experimental embryology might permit a decision between the above two hypotheses. This suggests that no conclusion is now possible. It is, nevertheless, interesting that the mesonephritic origin of the interrenal Anlage is claimed only by authors who studied saurian embryos (*Lacerta muralis*, *Lacerta agilis*, *Anguis fragilis*); no embryological work on the interrenal tissue of Testudines, Crocodilia, or Ophidia mentions this mode of ontogeny. Also, the last claim for a mesonephritic origin of the interrenal tissue is much less strongly stated than are the older studies; the possibility of a *mixed* origin, involving both peritoneal and mesonephritic tissues, is not excluded.

Start of Interrenal Activity

I know of no papers presenting physiological data concerning the start of activity in the embryonic interrenal tissue, and the morphological studies provide no more than a first approximation. Indeed, the only technique other than descriptive microscopical anatomy that has been applied to the embryology of the reptilian adrenal gland is the identification of lipid globules revealed by their osmiophilic reaction or by staining with dyes of the Sudan series. No workers seem to have attempted specific tests for steroids or for $^{\Delta5}$-3-β-hydroxysteroido-dehydrogenase.

Particulate lipids appear in the interrenal cells of saurians in a relatively early embryonic stage. The interrenal cells of *Lacerta* contain lipids as soon as the Anlage has fused into a continuous interrenal cord, while 120 to 150 mg embryos of *Anguis fragilis* show particulate lipids in the interrenal cells. Even 3 mm embryos of *Xantusia* contain osmiophilic lipids.

Embryonic Development and Start of Activity in the Adrenal Tissue

Opinions differ about the origin of reptilian interrenal tissue, but there is a complete agreement that reptilian adrenal tissue has the same origin as that of other vertebrates and that this tissue forms and begins to function relatively late in ontogeny.

Earlier authors noted the presence of small clusters of sympathopheochromoblasts in the connective tissue between the notochord, aorta and mesonephros of reptilian embryos. These cells were identified by the intense colouration of their cytoplasm, but did not yet yield a chromaffin reaction, specific for polyphenolic components. The chromaffin cells begin to differentiate after the interrenal tissue is clearly differentiated and already contains particulate lipids; the contact between the interrenal and adrenal tissue develops only in late ontogenetic stages, or after hatching.

Modern investigations confirm these results entirely. Bimmer (1950) shows that the differentiation of the adrenal tissue of *Lacerta* and *Chalcides* embryos occurs in an area that extends considerably posterior to the interrenal tissue and that the chromaffin reaction only occurs just before hatching. Adrenal cells of *Xantusia vigilis* first appear in 3 mm embryos. At this time the interrenal tissue still contains osmiophilic lipids; the time of contact for the two tissues was noted above in the description of interrenal development. Cytological signs of secretion only appear with the approach of hatching, and the topographic relations between adrenal and interrenal tissues become fixed only after the first months of life.

Macroscopic Anatomy

The first investigators noted the great variability of form and size in reptilian adrenal glands, and all recent reviews emphasize this. Its topographic position tends to vary with the systematic position of the animals studied, but certain of its relations are relatively constant.

Quantitative data concerning the size of the adrenals are rare and cannot easily be compared because of such parameters as the sex and age of the animals, as well as their stage in the annual cycle. The lack of homogeneity probably magnifies the inherent variability. In general the two adrenals weigh between 10 and 60 mg for each 100 g of body weight.

Table 5.1. Some data on the relative weight of reptilian adrenal glands

Species	Relative weight of the adrenal (mg/100g)
Caretta caretta	33
Emys orbicularis	10
Geochelone chilensis	9.5
Testudo sp.	10
Heloderma suspectum	24.96
Lacerta viridis	40
Tupinambis teguixin	33.5
Varanus salvator	14
Coluber sp.	40
Natrix natrix	50
Thamnodynastes sp.	24.15-55.6
Tomodon sp.	15.43-34.2
Caiman latirostris	19

The colour of the adrenal gland also varies from light yellow to reddish, not only from one species to another but also during the annual cycle of a single species. This colour is produced, at least in part, by the relative abundance of particulate lipids in the interrenal cells and by the amount of blood in the organ.

The shape and the relations of the adrenal with adjoining organs are quite variable and differ with the major systematic groups, so that they must be treated separately.

The adrenal gland of turtles is dorsoventrally flattened, has more or less regular borders, and lies against the kidney. The dorsal limit of the gland is difficult to see, since only the basement membranes of the cellular strands separate the gland from the renal parenchyma. The islets of interrenal and adrenal tissue regularly penetrate a short distance into the parenchyma. The ventral sides of the glands are covered by the parietal folds of the peritoneum. A peritoneal fold, attached to the ventrolateral side of the adrenal gland, ensheathes each gonad and the proximal part of each gonoduct to form the mesorchium of the male and the mesovarium of the female. The gland

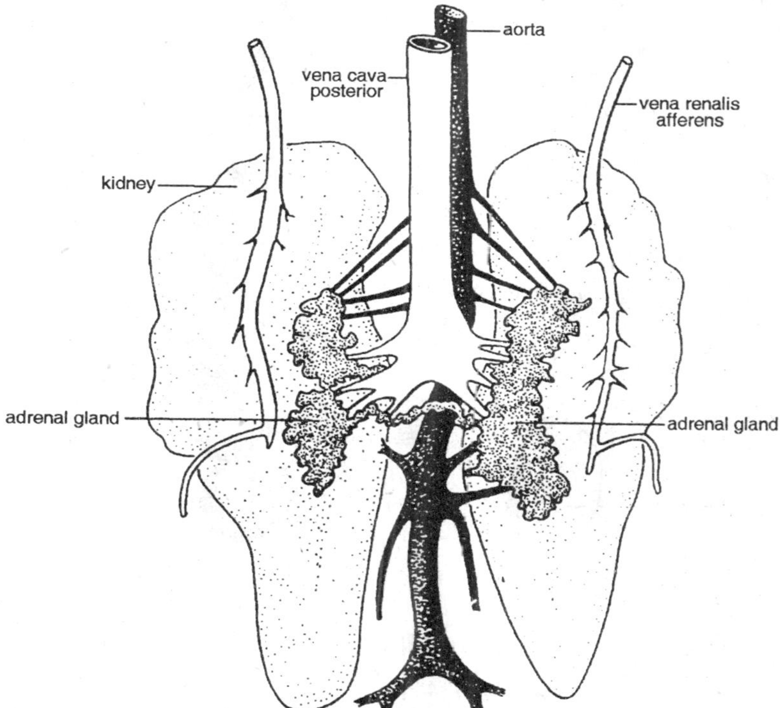

Fig. 5.1. Location of the adrenal glands at the ventral surface of the kidney in Pseudemys scripta.

also lies close to the kidney and the proximal segments of the genital system and is crossed by the efferent renal veins.

The adrenal gland of *Sphenodon punctatus*, the only Recent rhynchocephalian, lacks a direct relation with the kidney. Its topographical relations with the genital system are even closer than in turtles, since the adrenal is entirely incorporated in the mesorchium of the male or mesovarium of the female. It is impossible to separate the adrenal gland of the male from the epididymis or that of the female from the Wolffian vestiges, since a common sheath of connective

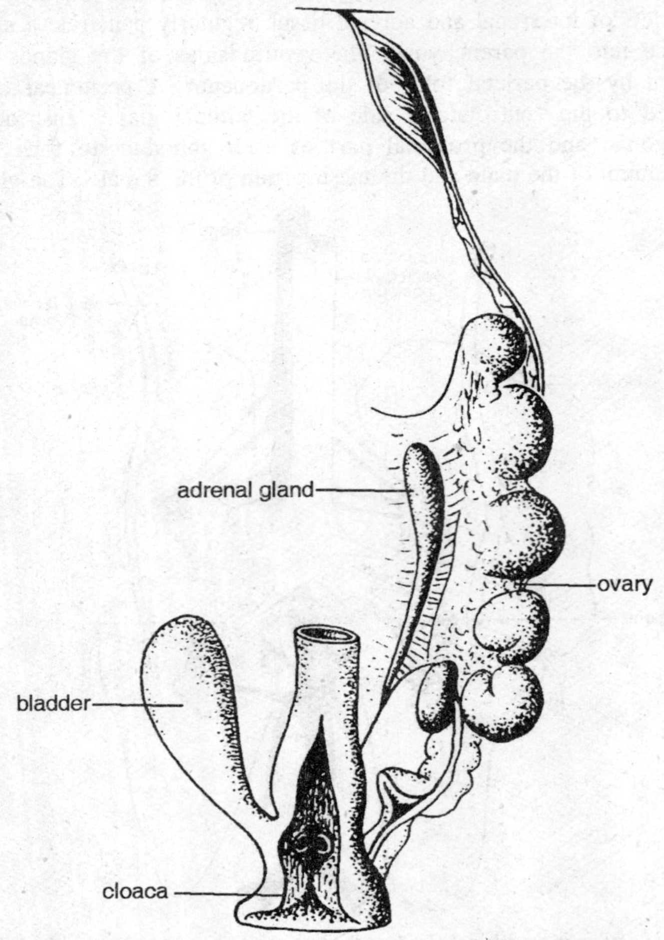

Fig. 5.2. Diagrammatic representation of the adrenal gland in a Sphenodon punctatus.

tissue covers both. The anteroposterior extension of the organ is considerable, with the anterior pole applied to the gonad and the posterior pole terminating some distance from the kidney. The transverse diameter of the adrenal is, on the contrary, relatively small, so that the organ has the shape of an elongate spindle.

The adrenal gland of the Squamata is also incorporated in the mesorchium or mesovarium of the gonad. The gland's relations with

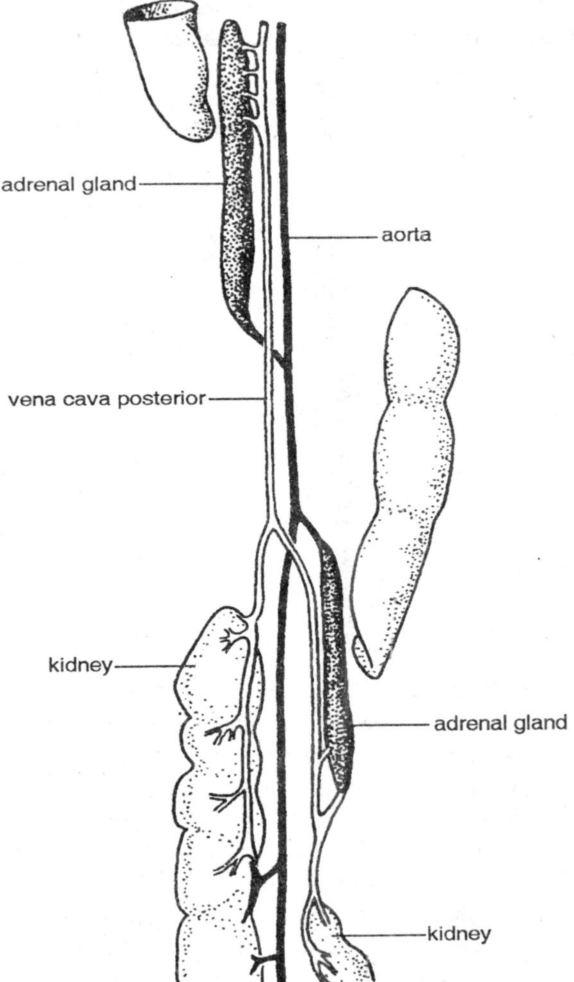

Fig. 5.3. Location of the adrenal gland in Thamnophis sirtalis.

the testes or ovaries and the gonoducts are very similar to those noted in *Sphenodon*. The transverse diameter of the gland is generally relatively small. Some lizards have globular or cylindric adrenal glands, but in snakes the adrenal is pronouncedly elongate and filiform. Its relations to the mesonephric derivatives are as close as in *Sphenodon*.

The adrenal gland of crocodilians is entirely retroperitoneal; the rather massive cylindric organ lies dorsal to the gonad and lateral to the genital duct. Its posterior pole is insinuated into the parietal peritoneum along the ventral side of the kidney.

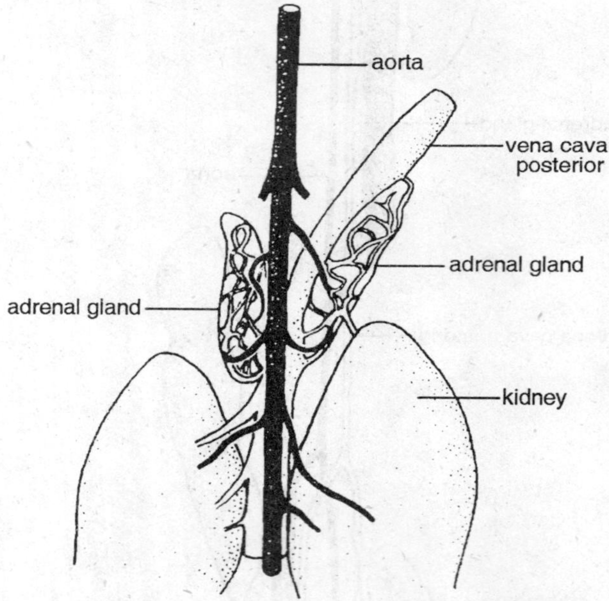

Fig. 5.4. Location of the adrenal gland in Alligator mississippiensis.

In spite of the differences in the form of their adrenals and in their intramesenteric position in lepidosaurs and in contrast to their retroperitoneal position in turtles and crocodilians, all reptilian adrenal glands share several common macroscopic characteristics. Thus they always lie dorsal to the gonads and in close proximity to the genital ducts, the latter being medial or ventral to the gland.

The adrenal glands of all reptiles except turtles are markedly asymmetric. The right gland lies well anterior to the left so that its anterior pole may, in certain Squamata, approach the posterior part of the liver.

The blood supply of the reptilian adrenal gland was shown by Gratiolet (1853) to contain a venous portal system, so that it receives venous as well as arterial blood. This peculiarity has been confirmed by all subsequent workers.

The arterial blood reaches the adrenal gland via branches of the dorsal aorta, genital arteries (testicular in males, ovarian in females), and renal arteries. The relative importance of the different blood vessels is group specific. The works of Pettit (1896), Beddard (1904a, 1904b, 1904c, 1904d, 1906a, 1906b), Spanner (1929) and Hebard and Charipper (1955), as well as other papers cited in the general reviews mentioned in the introduction, show that in turtles arterial blood is supplied via the renal arteries. Branches of the dorsal aorta are most important in the arterial irrigation of crocodilian adrenal glands, while branches from the dorsal aorta and genital arteries furnish arterial blood to lepidosaurian adrenal glands. The aortic supply is much more important in lizards than in snakes. One special case is worth mentioning: in *Gerrhonotus multicarinatus* branches of the several large arteries in the vicinity, including the mesenteric arteries contact the adrenal gland and ramify on it.

The afferent vessels of the adrenal portal system are of a venous type, and their origin varies with the systematic position of the species. In turtles the three, very short, afferent veins of this portal system arise from the renal portal veins. The afferent adrenal vein of *Sphenodon punctatus* arises from the anterior part of the kidney and, in adult specimens, follows the derivatives of the Wolffian body to reach the adrenal gland and ramify on it. The trunk of the afferent vein also receives parietal and vertebral veins. Two afferent adrenal veins occur in snakes and both arise from the dorsal body wall. The adrenal portal system varies greatly in lizards. Some forms have only one afferent adrenal vein that passes from the dorsal wall of the body, touches the posterior extremity of the adrenal gland, and ramifies on it (*Phrynosoma cornutum*). Other species have two afferent adrenal veins passing from the dorsal body wall (*Agama agama*, *Lacerta viridis*), and in still other species the number of afferent adrenal veins varies from one to three. In crocodilians one vein, arising from the dorsal body wall, forms the afferent portion of the adrenal portal system.

All the blood from the adrenal gland, whether derived from the adrenal arteries or from the afferent veins of the portal system, drains to the vena cava posterior. The pattern of the venous return is determined by the relations of the adrenal gland to the vena cava, which is formed by the junction of the efferent renal veins. In turtles, the latter may receive the adrenal drainage while they cross the middle of the adrenal parenchyma. Distinct efferent adrenal veins join the spermatic veins in the male of *Sphenodon punctatus*; the homologous veins of the females have not been studied. In squamates and crocodilians, the venous blood flows directly from the right adrenal gland into the vena cava posterior, while the left gland is drained by veins entering the renal vein.

Less precise information is available about the innervation of the reptilian adrenal gland than about its vascularization. Most authors report that the nerves to the organ originate in the lateral sympathetic ganglia of the paravertebral chain, but the structure seems to be very variable, especially in the number of ganglia which participate in this innervation. Clusters of ganglion cells have, furthermore, been reported in the immediate proximity of the adrenal gland or in contact with it in all reptiles thus far studied, but the origin and terminations of the nerve fibers to the organ have never been studied systematically. Hebard and Charipper (1955) mention the very rich innervation of the glandular parenchyma, but give no information about the extraglandular course of these fibers.

Microscopic Anatomy

Comparison of the descriptions in the classic papers of Braun (1882), Vincent (1896), Minervini (1904), and Poll (1906) discloses the fundamental similarity of the adrenal histology in all reptiles and confirms its morphological resemblance to that of birds. However, noticeable differences do exist between representatives of different orders of reptiles. These differences are sometimes reflected in the histology of the organ, while in other cases they are in the relative abundance of connective tissue and glandular parenchyma, in the distribution of interrenal and adrenal tissues, or in the presence or absence of chromatophores and neurons in the gland. Hence it is necessary to consider the major groups separately.

Testudines

As noted under macroscopic anatomy, the adrenal gland of Testudines is not strictly delimited dorsally, and renal and adrenal parenchyma may intermingle there. Nodules of interrenal or adrenal tissue may be scattered along the walls of efferent renal veins at the ventral side of the gland.

A connective tissue capsule of variable thickness covers the ventral face of the gland. Dorsally, this sheath is represented only by capsular expansions, extending from the ventral and lateral faces of the gland and limiting the cellular cords. The development of the ventral wall of the capsule is proportional to the size of the animal; the capsule is thin in *Testudo graeca*, well developed in *Clemmys caspica leprosa*, and very thick in *Lepidochelys olivacea*. In *Caretta caretta* the connective tissue capsule forms an uninterrupted plate bridging the median line and resulting in a median coalescence of the two adrenal glands.

The capsule is formed mainly of randomly oriented, collagenous fibers. Hebard and Charipper (1955) claim that the fibers are thickest in the external zones of the capsule, where it connects to the neighbouring connective tissue. The internal zones, in contact with the parenchyma, contain thinner collagenous and reticular fibers. Elastic fibers are rare in the capsule, but frequent in the walls of blood vessels crossing it. Most cells of the connective tissues are fibrocytes; mast cells are rather rare. The texture of the gland is produced by expansions of the internal layers of the capsule that contain rather thin collagenous and reticular, but no elastic fibers.

Within this capsule the interrenal tissue forms a net of anastomosed cords. Each cord is composed of radially arranged cells. One pole of an interrenal cell contacts an adjoining cell, while the opposite pole contacts the capsular expansion that separates the cord from the blood vessels. Consequently a cross section passing exactly along the longitudinal axis of a cord shows two rows of cells. The orientation of cords in the organ is variable, but the cellular arrangement is very uniform and rarely shows regional differentiation. In *Chrysemys picta* the peripheral cords of the adrenal contain narrower and more elongate cells than do the central cords.

The adrenal tissue of the Testudines consists of cellular clusters distributed irregularly in the gland. Some of these clusters are in contact with the connective tissue capsule; other, larger ones, lie in the middle of the parenchyma. Capsular expansions, comparable to those which delimit the interrenal cords, cover the more or less globular, ovoid, or irregularly shaped clusters of adrenal cells. In contrast to the condition frequently observed in lepidosaurians, the clusters do not expand between the interrenal cords.

There seem to be no quantitative data about the relative volumes of the two tissues in turtles; subjective statements, of course, represent only a first approximation. Volumetric differences may occur, since Hebard and Charipper (1955) note the particular abundance of adrenal tissue in the gland of *Clemmys guttata*.

Neurons are common in the adrenal glands of all Testudines studied. Small clusters of neurons are often included in the capsule or in the space between it and the glandular parenchyma. Isolated nerve cells or small groups of them are also found in the parenchyma; most are in contact with groups of adrenal cells. Nerve fibers, which are often quite large, may be seen amid the clusters of adrenal cells, even when these are only stained for topography. Neurofibrillar impregnations show an abundance of nerve fibers along the capsular expansions and neuronal penetration into the clusters of adrenal cells. Penetration of nerve fibers into the interrenal cords is much rarer.

Dendritic melanophores are found in the peritoneum, the connective tissue capsule, and even the parenchyma of the adrenal gland in some species. They are particularly noticeable in *Testudo graeca*.

Rhynchocephalia

The adrenal gland of *Sphenodon* lies within the gonadal mesentery and is consequently easier to recognize than that of Testudines.

A generally quite thin connective tissue capsule containing collagenous and reticular fibers totally ensheathes the gland. Since the adrenal gland lies adjacent to the vas deferens, the connective tissue sheaths of both organs are continuous. In females the capsule of the adrenal gland is less closely united to the connective tissue of the proximal segments of the genital system. As in other reptiles, partitions, formed of very slender collagenous and reticular fibers and projecting inward from this capsule, subdivide the organ and form its supporting framework. The thinness of the capsule explains the impossibility of determining whether it has both external and internal layers like those of turtles.

The interrenal cords anastomose and are, as in other reptiles, constituted of radially arranged cells. One pole of each cell is applied to the capsular expansion that separates it from the blood vessels. The orientation of the cords is irregular within the parenchyma, and no regional differences occur.

The adrenal gland of *Sphenodon punctatus* differs from those of all other reptiles in the disposition of the adrenal tissue. The interrenal cords are dorsally covered by a relatively thick, ventrally concave, trough-shaped mass of adrenal tissue. As in other lepidosaurs, expansions from this mass penetrate between the interrenal cords into the interior of the organ. Variably sized islets of adrenal cells occur within the parenchyma. The adrenal gland of *Sphenodon* is fundamentally distinct by having more or less voluminous clusters of this tissue on its ventral surface, although this surface never contains adrenal tissue in any other lepidosaur. Only in the lateral regions of the gland and at certain points in the ventral zone do the interrenal cords touch the capsule.

The accumulations of nerve cells are rather frequent near the dorsal surface of the adrenal gland, but they are smaller than in Testudines. These cell groups either contact the capsule or lie in the dorsal layer of the adrenal tissue. The much rarer nerve cells embedded in the parenchyma are often isolated. In cross sections stained for topography, the nerve fibers seem to be much rarer and thinner than those in the adrenal glands of turtles.

Dendritic melanophores occur in the capsule as well as in the parenchyma of the adrenal gland. They are rarer than in *Testudo* as are mast cells.

Squamata

The adrenal gland of the Squamata lies within the gonadal mesentery near the gonad and the proximal parts of the gonoducts, as does that of *Sphenodon punctatus*. The right gland is attached to the vena cava posterior.

The connective tissue capsule, again formed of collagenous and reticular fibers, is generally thicker in the larger glands. The fibrous constituents of the connective tissue capsule are not oriented. Smaller glands lack zonation in the capsule; larger ones have the thicker collagenous fibers externally and the reticular and thinner collagenous fibers located internally in contact with the parenchyma. Connective tissue cells morphologically similar to fibrocytes are rare. Dendritic melanophores have been found in only a few lizards (especially in the Diploglossa). Mast cells are very rare.

As in other reptiles, capsular expansions, of collagenous and reticular fibers, partition the glandular parenchyma.

The interrenal cells are arranged in cords as in other reptiles. The cords are oriented randomly and show no regional differentiation in structure. Hebard and Charipper (1955) recognize three patterns of cells within the interrenal cords of lizards. The first is found in the Gekkota and Scincomorpha, including the Gekkonidae, Dibamidae, Pygopodidae, Xantusiidae and Scincidae, the second in the Diploglossa, and the third in the Iguania.

The essential morphological character of the first, gekkotan and scincomorphan type of interrenal tissue is a radial disposition of tall conical cells with the nuclei generally located towards the periphery of the cords. In the Diploglossa prismatic cells, lower than those of the Gekkota, are irregularly distributed and the cords lack regular rows of cells. The Iguania have regularly arranged double rows of interrenal cells with central nuclei.

Actually the position of the nuclei in the interrenal cells is so variable, even in a single individual, that it cannot be utilized as a criterion. The variations in the position of the nuclei presumably correspond to stages in the secretory cycle and are discussed below.

The adrenal tissue of all Squamata thus far studied consists of a compact dorsal layer from which extensions penetrate into the parenchyma, and of isolated islets that lie within the parenchyma on the border of the interrenal cords in immediate proximity to the blood sinuses.

The thickness of the dorsal layer varies in different regions of the gland. Its center is well developed, while its edges are thinner in most snakes; the layer is thick anteriorly and becomes thin posteriorly in Scincomorpha. The layer is particularly well developed in amphisbaenians, but very poorly so in Leptotyphlopidae. Similarly the extensions derived from the dorsal layer of the adrenal tissue differ in their development in different groups. The snakes, Diploglossa, Gekkota and Iguania are characterized by rather short but thick extensions of adrenal tissue which do not deeply penetrate the interrenal tissue, while the extensions of the Scincomorpha are longer and thinner. In amphisbaenians the extensions are so numerous that parasaggital sections, far from the median plane, seem to show complete mixing of the two tissues like that in crocodilians and birds.

The number and size of the islets of adrenal tissue included within the parenchyma and lacking connections with the dorsal layer also vary. Quantitative studies are rare, and even when the relative volumes of the interrenal and the adrenal tissue are given, no distinction is made between the dorsal layer and the isolated islets. The latter are generally numerous but small in the Squamata. Clusters of adrenal cells as large as those found in turtles are very rare.

In squamates, nerve cells, isolated or grouped into small ganglia, almost always occur at the dorsal face of the adrenal gland, either in contact with the connective tissue capsule or in the dorsal mass of the adrenal tissue. Ganglia rarely occur within the parenchyma, but isolated nerve cells, which are almost always in contact with islets of adrenal cells, are common.

In Diploglossa the parenchyma of the adrenal gland contains only few dendritic melanophores and pigmented macrophages histologically similar to hepatic and splenic macrophages. Mast cells are also rare.

Crocodilia

The adrenal gland of crocodilians is sharply defined; not only is its anterior part distinct from the adjoining gonad and genital ducts, but its posterior end is also clearly separated from the ventral face of the kidney. Histological investigations on *Crocodylus niloticus*, *Alligator mississippiensis* and *Caiman crocodilus* prove that the structure of the organ is very uniform within this order.

The connective tissue capsule, which completely ensheathes the adrenal, is relatively thicker than that of other reptiles. Its external layers, formed of bundles of thick collagenous and reticular fibers, send sheets inward to partition the parenchyma. Only the tunics covering blood vessels crossing this capsule contain elastic and muscular fibers.

The interrenal cords resemble those of other reptiles and touch the capsule in some places. In other regions the cords are separated from the capsule by a discontinuous layer of adrenal tissue that sends extensions to the center of the organ, so that the two tissues are closely intermingled throughout the adrenal gland. This arrangement is exactly like that noted in birds.

Nerve cells, isolated or grouped into small ganglia, frequently lie in contact with the dorsal face of the gland, but they generally remain outside the capsule. Crocodilians have fewer nerve cells in the parenchyma than do any other reptiles.

Structural types of Adrenal Glands

Three structural arrangements of the adrenal tissue and interrenal cords may be distinguished in sauropsids.

In the first, found in turtles, the adrenal tissue is dispersed in irregular or ovoid clusters. These clusters are randomly spread throughout the organ. Different species of turtles contain variable quantities of adrenal tissue associated with the interrenal cords in both the periphery and the center of the adrenal gland.

The second type, found in lepidosaurs, is characterized by a clear tendency for aggregation of the adrenal tissue. In *Sphenodon punctatus* much of the tissue is arranged in a dorsal layer which covers the mass of interrenal cords, while isolated islets of adrenal tissue occur within the parenchyma and on the ventral face of the gland. The pattern seems to be the same in squamates, but there the concentration is more pronounced, and the isolated islets of adrenal tissue have practically disappeared from the ventral face of the adrenal gland. Most of the adrenal tissue is located in the dorsal layer, and the total volume of the extensions and of the isolated islets within the parenchyma is small.

The third adrenal type, characteristic of archosaurs, is found in crocodilians and birds. Almost all the adrenal tissue is arranged in bands which, in cross section, seem to alternate with the interrenal cords. Histological sections of crocodilian adrenal glands do not indicate a zonation, although those of certain birds do, uniquely, show indications of such a zonation.

The microscopic anatomy of the sauropsid adrenal gland thus supports the accepted subdivisions of the reptiles and the present conceptions of their phyletic relations.

Aberrant Clusters of Interrenal Tissue and of Chromaffin Cells

As already noted, the adrenal tissue does not include all the chromaffin elements, either of vertebrates in general or of reptiles in particular, nor does interrenal tissue occur only in the adrenal gland. Knowledge of their embryology explains the frequent cases of aberrant nodules of interrenal or adrenal tissue.

No systematic inventory of the aberrant sites of these two reptilian tissues has yet been compiled though some incidental observations are reported by Poll (1906). Such nodules are particularly common dorsal to the adrenal gland *sensu strictu*, in the connective tissue that surrounds blood vessels of this region, and in the paravertebral connective tissue. In turtles and crocodilians even the kidney may contain islets of adrenal or interrenal tissue that have lost contact with the body of the gland.

These islets generally represent only a relatively small mass of tissue superficially of little importance. However, their presence must be considered by physiologists, especially when interpreting the consequences of surgical adrenalectomy.

Cytological and Histochemical Characteristics of Interrenal Cells

Though the adrenal gland of reptiles is anatomically very diverse, the cytological and histochemical characteristics of its parenchymal elements are remarkably uniform. Hence only a general description seems necessary. Modifications observed in some species are mentioned in the course of the description.

The interrenal cells of all reptiles thus far studied are generally prismatic. Their mean height varies from 15 to 30 μ; size variations in different species cannot be easily interpreted because the interrenal cells may vary in size within one animal. The basal pole of each cell is in contact with the capsular expansion that separates it from a blood vessel; the apical pole meets another interrenal cell. The position of the nucleus in the cells is similarly variable. Though older works generally report that it lies in basal position, near the capillaries, this is not universally true. Gabe et al. (1964) have shown that the nucleus often lies near the capillary pole in the Scincomorpha, yet in all reptiles certain interrenal cords show obviously central nuclei. Detailed analysis of the sections proves that the position of the nuclei varies considerably in different cords of the same adrenal. The nuclei occasionally lie in an apical position with the basal pole occupied by lipidic droplets, as described for the interrenal cords of *Thamnophis sirtalis*. It is highly probable that the several different positions of the nuclei correspond to functional stages of the cells, but the available histophysiological data do not yet permit unequivocal interpretations.

The nuclei also vary in form and structure. Extreme variants include nuclei of oval or circular cross section with rather small, diffuse, irregularly distributed blocks of chromatin, and crumpled nuclei with irregular contours and dense chromatin. The nucleoli are very easy to recognize in the first case and difficult to distinguish in the second.

The structural differences in the nuclei correlate directly with the abundance of particulate lipids in the cytoplasm. Cells with clear nuclei and a well defined nucleolus are poor in particulate lipids, while cells with a crumpled nucleus contain great quantities of lipids. Cross sections suggest that the pressure of the lipid inclusions has deformed the nuclei.

The hyaloplasm of the interrenal cells is feebly acidophilic; the general colour is very variable and depends on the abundance of the particulate lipids which are dissolved during paraffin embedding. If the inclusions of the hyaloplasm are abundant, the cross sections show only a fine, slightly coloured cytoplasmic net, with gaps corresponding to the lipid inclusions. When the cell is poorer in particulate lipids the hyaloplasm will, on the contrary, be well developed and its acidophily stronger. The ultrastructure of the chondriome of reptilian interrenal cells is yet unknown, though this organelle has been described from fresh tissue, and both Janus Green vital staining and the classical mitochondrial techniques have been applied. All data agree about the form of the chondriome, represented by rather short chondrioconts and by mitochondria. All these organelles are concentrated in thin cytoplasmic pillars in cells that are rich in lipids. A perinuclear location of the chondriome has been reported in *Alligator mississippiensis*, and in *Trionyx sinensis*. No particular orientation of the chondrioma has been described in other species. The chondrioma of the interrenal cells are more easily demonstrated when particulate lipids are scarce.

Little is known about the Golgi apparatus of the interrenal cells and its ultrastructure has not been described. Scattered dictyosomes occur in the interrenal cells of *Gerrhonotus multicarinatus* and there is no "Golgi zone". Liu and Maneely (1959), on the other hand, report a Golgi apparatus at one pole of the nucleus of the interrenal cells of *Trionyx sinensis*. Light microscopy cannot show ergastoplasm in the interrenal cells of reptiles, and ribonucleic acids are demonstrable only in the nucleoli.

The interrenal cells are quite uniform in structure and show few species specific differences. Cells of a special form have been reported among the Squamata only in the Diploglossa. These diploglossan cells are as high as the ordinary prismatic cells among which they are interspersed, but their conical rather than prismatic shape permits rapid recognition of the adrenal glands of diploglossan lizards. The bases of these elements face the centers of the cords, and their apices are in contact with the capsular extensions. The cells do not otherwise differ in histology nor in histochemical reactions.

Conical cells have been observed in *Anguis fragilis*, *Ophisaurus koellikeri*, *Gerrhonotus multicarinatus*, *Anniella pulchra*, *Heloderma suspectum*, *H. horridum*, *Varanus griseus*, and *V. niloticus*. No other reptiles are known to have any interrenal cells of this type. Histological study thus confirms remarkably and unexpectedly the current concept of the systematic position of the Diploglossa which is, of course, derived from arguments of a totally different kind.

There are numerous recent studies of the histochemistry of reptilian interrenal cells. Glycogen has been found in Squamata, Testudines, *Crocodylus niloticus*, and *Sphenodon punctatus*. The abundance of this polysaccharide differs in different species and in different regions of the gland, but it is never very great. The interrenal tissue of the Squamata and the Testudines is generally richer in glycogen than is that of *Crocodylus niloticus*, while *Sphenodon punctatus* has the most glycogen of all the reptiles studied; indeed, glycogen is extremely abundant in many organs of *Sphenodon*. The amount of glycogen clearly varies during the annual cycle.

The first report of ascorbic acid in reptilian interrenal tissue is that of Knab (1942) for *Lacerta agilis*, *L. muralis*, and *L. viridis*.

Knab emphasizes the high individual variability, the absence of any preferential localization in the cell, and the low rate of enrichment of the interrenal tissue after subcutaneous injections of ascorbic acid in *Lacerta agilis*, *L. muralis* and *L. viridis*. This absence of a preferential localization has been confirmed in *Agama agama* and *Natrix natrix* and various squamates and three species of turtles. Only Liu and Maneely report a "Golgian" localization of the ascorbic acid in the interrenal cells. No one has yet checked for ascorbic acid in the interrenal tissue of the crocodilians or *Sphenodon*. No histochemically detectable acid mucosubstances occur in reptilian interrenal cells.

Histochemically detectable lipids are common in the interrenal tissue of reptiles and have been known since the classic works of the beginning of the twentieth century. The lipids from droplets or larger plaques which are osmiophilic and stain with all the usual lipid dyes. Their histochemistry has been studied for representatives of all the reptilian orders of the class. Generally the neutral lipids stain rose with Nile blue sulfate (method of Lorrain Smith) and represent only a small fraction of the globular lipids; this observation explains the abundance of birefringent lipidic inclusions that produce the black cross phenomenon in polarized light. Variants of the reactions of Liebermann and of Windaus for cholesterides always yield positive results. The occurrence of lipidic carbonyles is shown by the positive results of pseudoplasmal reactions and by those of Ashbel and Seligman. Thus the lipids of the reptilian interrenal tissue agree in their histochemical characteristics with those of other vertebrates.

Hebard and Charipper (1955) describe three patterns of distribution of the lipidic inclusions in interrenal cells. In the first, lipidic droplets are found at the vascular pole of the cells, while the nuclei lie at the opposite pole, near the central part of the cord. This pattern has only been found in *Thamnophis sirtalis*.

The second situation is the reverse of the first, and the nuclei are then found at the vascular pole of the cell. This situation occurs in certain lizards, especially *Eumeces obsoletus*. In the third case, all parts of the cell contain lipidic inclusions, while the nuclei are central. This pattern occurs in all turtles and crocodilians and in some lizards.

This classification is obviously schematic. Only the relative frequency of the several cell types varies, since careful search of cross sections shows all three types in any one individual.

Besides particulate lipids, the interrenal tissue of different species and individuals contains variable amounts of oxidative derivatives of lipids. These granules belong to the group of the chromolipoids and, in all their histochemical properties, correspond to the lipofuscins of the reticular zone of the mammalian adrenal cortex. These inclusions are circular in cross-section, yellow in unstained sections, without melanins, and more frequent in turtles than in lepidosaurs and crocodilians.

Protids are rare in the reptilian interrenal cells, and there are no reports of accumulation of these compounds. Cytoplasmic ribonucleins have already been noted to be undetectable by light microscopy. The total mineral contents of the cells is quite low. Ionic iron is absent, which is remarkable since it is rather frequent in the reticular zone of the mammalian adrenal cortex.

The presence of a nonspecific alkaline phosphomonoesterase has been observed in representatives of all reptilian orders which is in accord with data for the mammalian adrenal cortex. Liu and Maneely (1959), however, could not demonstrate alkaline phosphatase in the interrenal tissue of *Trionyx sinensis*. The works cited earlier also document the absence of nonspecific acid phosphomonoesterase in reptilian, interrenal tissue. Strong $^{\Delta 5}$-3-β-hydroxysteroido-dehydrogenase activity has been noted in the interrenal tissue of *Emys orbicularis*, *Varanus niloticus* and *Natrix natrix*. Similar enzymatic activity occurs in the adrenal cortex of mammals and, more generally, in all tissues involved in active steroidogenesis.

A unique case of "colloidogenesis" of the interrenal tissue in an adult male *Gerrhonotus multicarinatus* is reported by Gabe et al. (1964). While none of the other organs proved anomalous in macroscopic or microscopic study, a great number of the interrenal cells of this animal were rich in spherical inclusions, clearly larger than the granules of chromolipoids. Acidophilic in paraffin embedded material and strongly PAS-positive, these inclusions closely resemble the adrenal "colloid" of mammals. Unfortunately it is not known whether this phenomenon is widespread.

The physiological observations on the zonation of the mammalian adrenal cortex explain why most recent authors consider the possibility of a similar zonation of the reptilian interrenal tissue. No physiological data suggest a functional specialization of different regions of the reptilian interrenal tissue, and there are only isolated and irregular indications of morphological zonation. Yet its existence cannot be disproven; the cells of the peripherical interrenal cords are distinctly smaller than those of the central cords in *Philodryas* sp and *Gerrhonotus multicarinatus*. Hebard and Charipper (1955) report differences in staining properties of the central and the peripheral interrenal cords of *Gerrhonotus coeruleus*. "Atrophic" cells in the peripherical interrenal cords of *Agama agama* and *Natrix natrix* show shrunken cytoplasm after paraffin embedding. Their "vacuoles", reflecting dissolved lipidic inclusions, are more numerous than those in the cells of the central cords. This observation is confirmed by the histochemical results. Especially in the Scincomorpha and Ophidia, the peripherical zones of the adrenal tissue are richer in particulate lipids than the central zones. Furthermore the $^{\Delta 5}$-3-β-hydroxysteroido-dehydrogenase activity is weaker in the peripheral than in the central interrenal cords of *Emys orbicularis*, *Varanus niloticus* and *Natrix natrix*. In spite of these incidental observations, the zonation of reptilian interrenal tissue is never so clearly established as is that of certain birds.

Cytological and Histochemical Characters of the Adrenal Cells

The adrenal cells, whose topographic position has already been noted in the section on microscopic anatomy, are generally smaller than the interrenal cells. When preparations are stained for topography, the cells may be distinguished by the stronger reaction of their cytoplasm, which is related to the absence or scarcity of particulate lipids and to the presence of cytoplasmic granules noted after fixation in most of the standard solutions. In contrast, the cytoplasmic granules of the mammalian adrenomedullary cells are only shown well when the gland is fixed in solutions containing potassium bichromate.

Certain histological characters are common to all reptilian adrenal cells, as well as to the chromaffin elements lying outside the adrenal gland proper. Other cytological and histochemical characters permit the division of adrenal cells into two groups, corresponding to the noradrenalin and the adrenalin cells recognized in the adrenal medulla of mammals.

Histological Characteristics Shared by the Two Types of Adrenal Cells

All isolated adrenal cells are oval or spherical in general shape. They become polyhedric when grouped in clusters. The cellular dimensions vary with the greatest diameter ranging from 12 to 18 μ. The nuclei are central, spherical, rather clear, and provided with one or more clearly visible nucleoli, even after a topographical staining. Acidophilic intranuclear inclusions, different from the nucleolus and visible under the light microscope, have been observed in squamate adrenal tissue. They are rarer in Testudines and exceptional in

Crocodylus niloticus. They appear to be absent in the adrenal tissue of *Sphenodon*. Electron microscopic observations would be essential for a correct interpretation of the relations of these structures with the intra-nuclear inclusions of mammalian adrenomedullary tissue.

There is insufficient ribonucleic acid in the cytoplasm of the adrenal cells to be demonstrated by light microscopy. The chondriomes are difficult to reveal because of the abundance of secretory granules of similar staining properties; the short chondrioconts and mitochondria lack clear orientation within the cytoplasm. The Golgi apparatus of reptilian adrenal cells has never been methodically studied with the electron microscope. In *Gerrhonotus multicarinatus* there are grouped dictyosomes, located at one of the poles of the nucleus of certain adrenal cells, in contrast to their diffuse location in interrenal cells. The silver impregnation of these structures by the method of DaFano seems easy. Liu and Maneely (1959) report a juxtanuclear net representing the Golgi apparatus in the adrenal cells of *Trionyx sinensis* fixed in osmium.

The secretory granules of the adrenal cells are strongly acidophilic when preserved in all the usual fixatives and take up hematoxylin lac. In crocodilians the latter reaction is the clearest. Numerous staining methods demonstrate the differences between the two cell types. It is quite understandable that different authors postulated the duality of the reptilian adrenal cells before study of other vertebrates suggested that these two types represent noradrenalin and adrenalin cells. This distinction is now definitely established and confirmed biochemically, while the ultrastructural characteristics of the secretory granules have also been determined.

Histological Characteristics of the Noradrenalin Cells

The size of reptilian noradrenalin cells varies within certain limits depending on their systematic position. In all orders except the Crocodilia the largest diameter approximates 12 μ. In *Crocodylus niloticus* the noradrenalin cells have a greatest diameter that reaches or surpasses 15 μ and are consequently larger than the adrenalin cells. This dimensional relation of the adrenal cells is quite unusual, since the noradrenalin cells are smaller than the adrenalin cells, not only in other reptiles, but also in all other classes of vertebrates.

The noradrenalin cells have large secretory granules dispersed throughout their cytoplasm. They are easily recognized, even under low magnification and correspond well to the definition of "rhagiochrome" cells. The secretory granules are acidophilic with all trichrome stains and take up ferric hematoxylin. They are less abundant than the cytoplasmic granules of the adrenalin cells and thus may be the cells referred to as "light". Evidently the term refers to the general colouration of the cell rather than the staining intensity of the individual granules.

The most obvious histochemical characteristic of reptilian noradrenalin cells is the strongly positive classical pheochrome (chromaffin) reaction. Treatment with liquids containing potassium bichromate stains the granules of these cells a more or less intense brown, though they maintain the capacity to react to P.A.S. and to ferric ferricyanide. The granules of the noradrenalin cells develop a dark brown colour after Hillarp and Hokfelt's (1955) methods with potassium iodate. After fixation in formaldehyde they show a white fluorescence, similar to that described for mammals, yet retain the capacity for azo and argentaffin reactions. The pattern thus resembles that well established for all other vertebrates.

The ultrastructure of the secretory granules of noradrenalin cells has only been studied in the snakes *Xenodon merremii*, *Bothrops alternatus* and *B. neuwiedi*. The granules may appear spherical, oval, polygonal, or comma-shaped with the polymorphism very characteristic of this type of cell. Each granule is surrounded by a very thin membrane which is difficult to fix. The average diameter of the granules is approximately 2000 Å, but the polymorphism produces considerable variation in size. The internal structure of the granules varies with different fixations, but their heterogeneity is always evident. Fixation by Caufield's method often produces "cristalline" structures in the granules which are evidently artifacts.

The cytoplasm of the noradrenalin cells is slightly acidophilic and lacks ribonucleic acids and particulate lipids, but demonstrates a variable quantity of glycogen. All reptilian adrenal cells studied thus far contain a small quantity of ascorbic acid.

The histoenzymological characteristics of the noradrenalin cells are identical to those of the adrenalin cells, so that the two types of cells cannot now be distinguished on the basis of their enzymatic activities. In contrast, the noradrenalin cells of the adrenal medulla of mammals can be detected by their feeble or nonexistent acid phosphomonoesterase activity, since the adrenalin cells show this strongly.

Histological Characteristics of the Adrenalin Cells

The adrenalin cells are larger (largest diameter approximately 18 μ) than the noradrenalin cells, except in the Crocodilia in which they are smaller (approximately 12 μ). Except for *Crocodylus niloticus*, the morphology of the secretory granules contained in the adrenalin cells resembles that described in the "hyalochrome" cells of the mammalian adrenal medulla. The granules are small, numerous, and very densely packed, giving the cytoplasm a homogenous appearance under low magnification. The abundance of secretory granules explains the stronger acidophilia in trichrome stains and why the adrenalin cells have been described as "dark cells". The secretory granules of the adrenalin cells of *Crocodylus niloticus* are as large as, but more numerous than, those of the noradrenalin cells.

The secretory granules give strong classical pheochrome reactions, but they are not revealed by the potassium iodate method of Hillarp and Hokfelt (1955) nor by that of Eranko (1955).

Wassermann and Tramezzani (1963) described the ultrastructure of the almost always regularly rounded secretory granules of the adrenalin cells. These granules have a diameter of approximately 1600 Å, and the variance is smaller than that of the noradrenalin granules. A space of low electron density separates a single membrane from the homogenous or finely granular center. The dimensions of this clear space vary according to the technique of fixation. No cristallization artifacts have been observed.

In most histochemical characteristics, the cytoplasm of the adrenalin cells resembles that of the noradrenalin cells. Both show equivalent amounts of glycogen and ascorbic acid and lack ribonucleic acids. In contrast to noradrenalin cells, adrenalin cells generally contain small quantities of particulate lipids, which are isotropic and "acidic" according to the present nomenclature. Chromolipids have also been reported, but their histophysiological significance has not yet been established. As noted above, the histoenzymological properties are indentical to those of the adrenalin cells. No alkaline-phosphatase activities are found in *Sphenodon punctatus*, Squamata, or *Crocodylus niloticus*. Gabe and Martoja (1962) found no reactivity in Testudines, but Liu and Maneely (1959) claim that the adrenal tissue of *Trionyx sinensis* has an alkaline phosphatase activity. The material should certainly be re-examined and the question settled.

The acid phosphatase activity of reptilian adrenal tissue is very strong. It permits quick detection of both groups of adrenal cells, but does not distinguish between them. No nonspecific esterase activities have been found, and the localization of acetylcholinesterase and butyrylcholinesterase corresponds to that of the acid phosphatases.

Distribution of the Two Types of Adrenal Cells

The orders of the reptiles clearly differ in the localization of the two types of tissues in the adrenal gland.

As noted in the section on microscopic anatomy, the clusters of adrenal tissue in the Testudines are randomly distributed. They may contain adrenalin cells, noradrenalin cells, or a mixture of the two types. Any of these three types of islets of adrenal tissue may be found at any point in the organ.

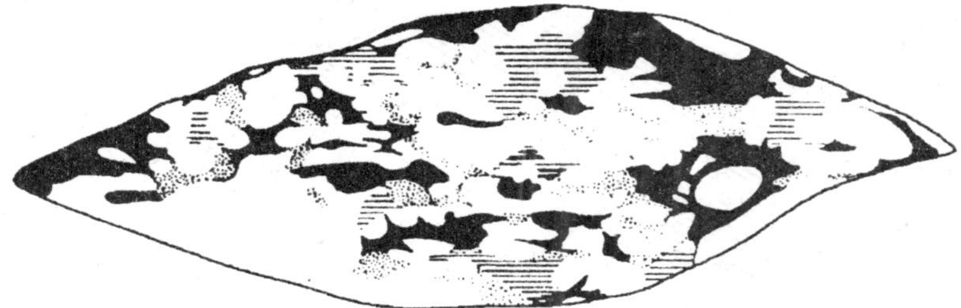

Fig. 5.5. Diagrammatic cross section through the adrenal gland of Testudo gracea.

The dorsal layer of adrenal tissue, which surrounds the interrenal cords of *Sphenodon punctatus*, contains primarily noradrenalin elements.

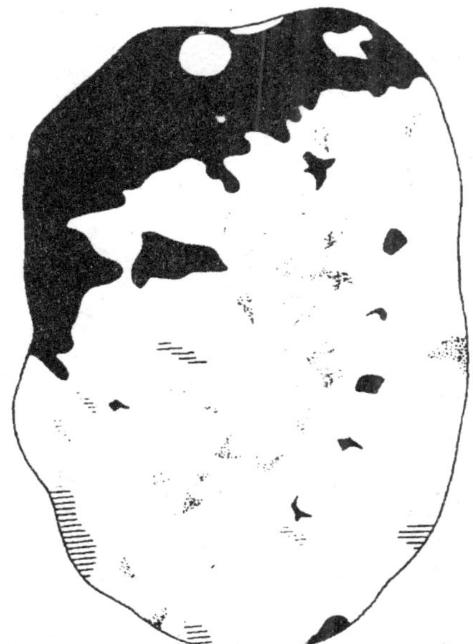

Fig. 5.6. Diagrammatic cross section through the adrenal gland of Sphenodon punctatus.

The ventral islets, which are in contact with the capsule of the organ, are either mixed or constituted exclusively of one of the two cellular types. The same is true for the islets included within the parenchyma.

The distinction between the two cellular types is much more pronounced in the Squamata. The dorsal layer of adrenal tissue consists exclusively of noradrenalin cells as do the digitations which penetrate between the internal cords. The clusters of adrenal tissue, incorporated in the parenchyma, contain only adrenalin cells.

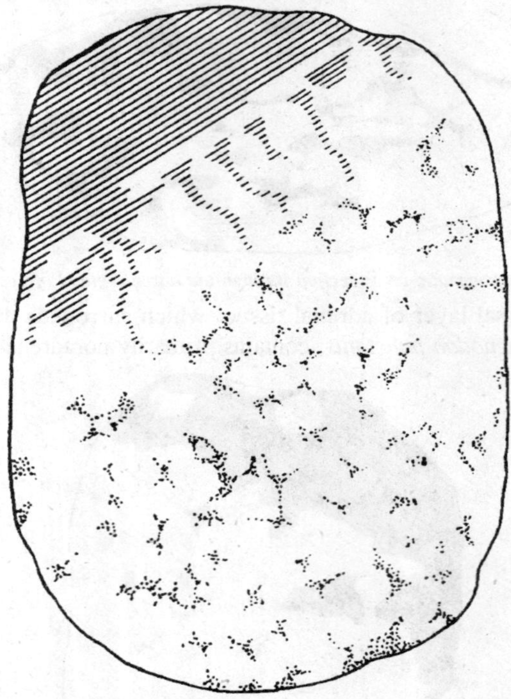

Fig. 5.7. Diagrammatic cross section through the adrenal gland of Vipera aspis.

This anatomical separation enabled Houssay and his co-workers to make a fluorometrical and chromatographical study of the two types of adrenal cells. Their results give a splendid confirmation of the histochemical data. In *Xenodon merremii* 96.77 ± 0.89 per cent of the catecholamines extracted from the peripheral layer of the adrenal appear to be noradrenalin; in contrast, 85 ± 3.12 per cent of the catecholamines extracted from the central part appear to be adrenalin. Chromatographic analysis indicates only a single spot, corresponding to noradrenalin, for extracts from the periphery, while extracts from the center indicate both a large spot corresponding to adrenalin and a slight spot corresponding to noradrenalin. Since noradrenalin is a normal precursor of adrenalin, these results do not contradict the morphological observation that the central part of the adrenal gland lacks noradrenalin cells.

In *Crocodylus niloticus*, all the cords of adrenal tissue are mingled with the interrenal cords and contain adrenalin as well as noradrenalin cells; the same situation is found in birds.

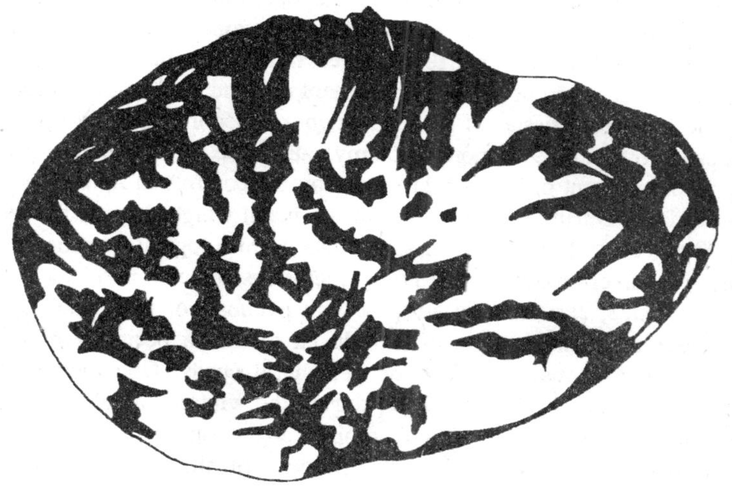

Fig. 5.8. Diagrammatic cross section through the adrenal gland of Crocodylus niloticus.

The cytological data, therefore, agree well with current ideas on reptilian classification. Turtles, lepidosaurs, and archosaurs show quite distinct patterns, while that of crocodilians resembles the pattern seen in birds.

Histophysiology

In the introduction I noted the meager utilization of reptiles in physiological studies. This explains the lack of data that would allow the definition of the histological modifications of the adrenal gland in different physiological states. Most of the results here presented concern the Squamata.

Effects of Hypophysectomy

Only the interrenal tissue shows modifications after hypophysectomy. Direct adenohypophysial control of the reptilian adrenal tissue of reptiles and the mammalian adrenal medulla is commonly lacking.

In *Thamnophis sirtalis* and *T. radix* hypophysectomy produces a significant diminution of the adrenal glands and a reduction of the volume of the interrenal cells. In *Xantusia vigilis* the amount of the lipidic droplets is reduced in part of the interrenal tissue after hypophysectomy. This decrease in lipids is accompanied by degenerative phenomena, such as nuclear pycnosis and a decrease in the size of the cells. During the weeks following the operation, the modified cells occur only in certain regions, while other zones in the same cross section show normal interrenal parenchyma. Hypophysectomy also induces a reduction of the cellular dimensions and of the lipidic droplets in the interrenal tissue of *Lacerta vivipara*; according to Wright and Chester Jones (1959) hypophysectomy causes the adrenal glands to lose up to 60 per cent of the weight shown by the controls in *Agama agama* in 30 days. During the same period the mean weight loss of the testis is 90 per cent. The adrenal tissue shows no histological modifications, but in the interrenal tissue the lipidic inclusions coalesce into "vacuoles" of great size, certain cells atrophy with pycnosis of their nuclei, and in some cases the cytoplasm is reduced to a thin perinuclear layer. These changes mainly affect the peripheral cords. One month after hypophysectomy of *Natrix natrix*, during the winter, the adrenal gland atrophies by 80 per cent, and similar degenerative phenomena are seen in *Agama agama*. The distinctness of the effects increases proportionally with the interval between surgical intervention and postmortem examination, but zones of normal interrenal tissue persist for at least 40 days after hypophysectomy.

Compensation for the Effects of Hypophysectomy

Data concerning the compensation for the atrophy produced by hypophysectomy are reported by Schaefer (1933) and Wright and Chester Jones (1957).

In two species of *Thamnophis* the atrophy disappears after implantation of hypophyses of the same species. Injections of mammalian ACTH produce the same effect. Actually the weight of the adrenal glands in *Natrix natrix* (mean weight approximately 60 g) is not modified by hypophysectomy if the animals receive daily injections of 0.5 I.U. of this hormone for the ten days preceding the autopsy. Histological examination confirms the normal state of the interrenal parenchyma, but often reveals an important thickening of the basement membrane.

Effects of ACTH Injections on the Adrenal Gland

The effects of ACTH injections on *Xantusia vigilis* and on *Natrix natrix* are similar. Both species react to small doses of ACTH by depletion of lipids in the interrenal cells. In *Xantusia* injection of higher doses produces degenerative modifications of the interrenal parenchyma, nuclear pycnosis, and a diminution of cellular size. After a prolonged treatment, the basement membrane of the interrenal cords of *Natrix natrix* shows the same thickening as is observed after the injection of ACTH into hypophysectomized snakes.

As Chester Jones (1957b) notes, interpretation of the degenerative modifications is not easy, since most of the commercial preparations of ACTH are contaminated by neurohypophysial hormones which might be responsible for some alteration of the corticotrophic effects. However, the reality of an effect due to high doses of ACTH, that is to a too strong stimulation of the interrenal parenchyma, cannot be excluded.

Injection of Corticosteroids

The only descriptions of corticosteroid effects are those of Miller (1952) and Wright and Chester Jones (1957). In *Xantusia vigilis* and *Natrix natrix*, the injection of desoxycorticosterone or of cortisone produces degenerative modifications of the interrenal tissue. These are indicated by a great increase of blood flow through the gland, a depletion of lipid in the cells, and even by atrophy which can be as pronounced as that following hypophysectomy. Cortisone is much more efficient than desoxycorticosterone in diminishing the size of the organ.

Changes in the Adrenal Gland during the Annual Cycle

Modifications of the adrenal gland during the annual cycle have been studied only in Squamata. Their interpretation is particularly difficult because few authors have associated histological examinations with physiological or biochemical tests which could have supported their conclusions.

Studies of such changes in the adrenal tissue are especially rare. Panigel (1956) claims that this tissue undergoes no notable modification during gestation in females of *Lacerta vivipara*, but his conclusions are based on preparations treated only with topographical staining methods. Techniques adequate for the study of chromaffin elements, especially the pheochrome reaction, clearly show inactivation of the adrenal cells during the first half of hibernation in *Vipera aspis*, *V. berus*, and *Anguis fragilis*. Signs of histological activity appear long before the end of the hibernation in these three species, and the strong histological indications of activity of the adrenal tissue correspond to a clear increase of glycemia. The onset of vernal activity coincides chronologically with a strong depletion of chromaffin granulations, the adrenal cells being clearly hypertrophied. The period of aestivation is accompanied by a diminution in size of the adrenal cells. The secretory granules are again abundant, but the modifications are much less pronounced than those accompanying winter involution. A new period of activity corresponds to the fall peak of sexual activity that precedes hibernation.

There are more studies of the modifications of the interrenal tissue, and these agree that the gland hypertrophies and the cell volume increases during sexual activity. Those authors who have studied the whole annual cycle note a decrease in the size of the interrenal cells at the start of the hibernation similar to that reported in the adrenal tissue. The nuclei become small and chromophilic, while the cytoplasm contains abundant lipidic inclusions, formed mainly by neutral lipids. As the gland enters its active phase, its size characteristically increases and there is a distinct lipidic depletion. Periods of sexual inactivity are marked by a new accumulation of particulate lipids in the interrenal cells. The validity of the histological criteria of the functional states of the adrenal gland has been demonstrated experimentally in mammals. In the latter, and presumably also in reptiles, the accumulation of particulate lipids in the interrenal cells is a sign of functional inactivity and lipidic depletion is a sign of activation. Furthermore, the activation of the interrenal tissue is paralleled by a decrease in the amount of glycogen within it. This suggests parallels to the comparable changes in the glycogen content of the adrenal cortex of mammals.

Conclusion

The adrenal gland of reptiles exhibits a pattern very similar to that known in other vertebrates; there are, nevertheless, some pecularities of this class.

Some doubt remains about the exact origin of the cells that form the interrenal cords. Most authors think that they derive from the coelomic epithelium; others that they derive from the wall of the Malpighian corpuscles of the mesonephros. The definitive histological characters of the interrenal cells appear at relatively early ontogenetic stages. The adrenal cells, which, according to all authors, have their origin in the sympatheticoblasts, only develop their definitive characters at a much later ontogenetic stage.

The adrenal gland of the Testudines resembles that of anuran amphibians in macroscopic anatomy, since it is closely adpressed to the ventral face of the kidney. In all other reptiles the relations of the gland with the metanephros are looser, and instead those with the gonad and gonoducts are closer. Similar relations also occur in birds.

At the microscopic level, the reptilian adrenal gland is characterized by the intermingling of the interrenal and the adrenal tissues. The modalities of this intermingling are different in the turtles, the lepidosaurs, and crocodilians.

All of the histological characters of the interrenal and the adrenal cells resemble those described in other vertebrates. Two types of adrenal cells have been found in all orders of reptiles, and their distribution within the adrenal gland is diagnostic. In spite of the paucity of experimental studies, this morphological similarity permits the application in reptiles of the histological criteria for the functional stages of the adrenal gland established in other vertebrates.

6

PANCREAS

A growing interest in comparative cytology and physiology has expanded our comprehension of reptilian pancreatic anatomy in the past decade, and recent contributions regarding the fine structure of the pancreatic islets of several species have been particularly prominent. However, the work of Thomas, published in 1942, remains the bulwark of our knowledge of reptilian pancreatic exocrine (acinar) tissue.

Very little has been published specifically pertinent to the gross anatomy or detailed features of the reptilian pancreas. Even Broman's account in Bolk's Handbuch (1937) offers little in this respect. Much more detailed studies of its structure and arrangement are needed, as is indeed true for other visceral organs of reptiles as well.

We have attempted to supplement this review with original observations of the gross anatomy, histology, and ultrastructure of the reptilian pancreas, using living specimens whenever possible. Thorough examination of every major group of reptiles in such perspective would require a lifetime. Reptilia as a group have suffered from lack of scientific attention, and the vast part of the attention they have received has been devoted to their bones and scales rather than to any of their viscera. The present account, accordingly, devotes greatest emphasis to those areas that we could review firsthand; the intervening areas will, therefore, appear incomplete reflecting the scanty relevant literature.

EMBRYOLOGY

Thoughtful discussion and extensive reviews of the literature on the origin and cytogenesis of reptilian pancreatic islet cells are found in the papers of Frye (1958; 1959), Bencosme (1955) and Bargmann

(1939), while the more general problems of pancreatic histogenesis are treated by Siwe (1926).

The pancreas in reptiles is usually derived from one dorsal and two ventral diverticula of the foregut which evaginate just anterior to the hepatic rudiments. While the dorsal pancreatic rudiment always develops, there is considerable specific variation in contributions to the definitive pancreas by the ventral Anlagen. Either both ventral diverticula contribute, or one or the other may be short-lived and disappear without forming part of the adult gland. These variations were discussed by Broman (1937). In the lizard *Xantusia vigilis* both ventral Anlagen appear, while only one, usually the left, contributes to the definitive pancreas. The late appearance of small islets in the anterior end of the pancreas, largely derived from the ventral Anlagen, suggests the possible existence of an islet contribution from that primordium on morphologic grounds alone.

Topography and Structure

Differences observed in the detailed gross anatomy of the pancreas in reptiles are largely dependent upon structural modifications of various portions of the gut, or relate to narrowing or broadening of the body cavity coincident with alterations in body form. Probably to a lesser, and as yet undetermined, extent, gross structural modifications of the pancreas may be phylogenetically explained.

Turtles

The gross form of the pancreas has been depicted in very few species of turtles. In *Pseudemy scripta*, the stomach and duodenum form a large "C"-shaped loop stretching across the wide center of the body cavity. Commencing at the pylorus and attached to the mesenteric border of the duodenum, the pancreas extends to the right and gradually enlarges as it approaches the right end of the latter organ; it follows the anterior, dorsal, and finally posterior duodenal loops, and terminates by leaving the duodenum and splaying out over the spleen, which is located more dorsally and usually more posteriorly than the terminal duodenal portion.

Anatomic arrangements in the soft-shelled turtle *Lissemys punctata* are similar to those in *P. scripta*. The pancreas lies along the mesenteric border of the wide duodenal loop, and a small part of the pancreas, connected by a thin shell of tissue to the mid-portion of the main body of the gland, is closely applied to the splenic capsule. The main body of the pancreas is vascularized by the pancreatic-duodenal

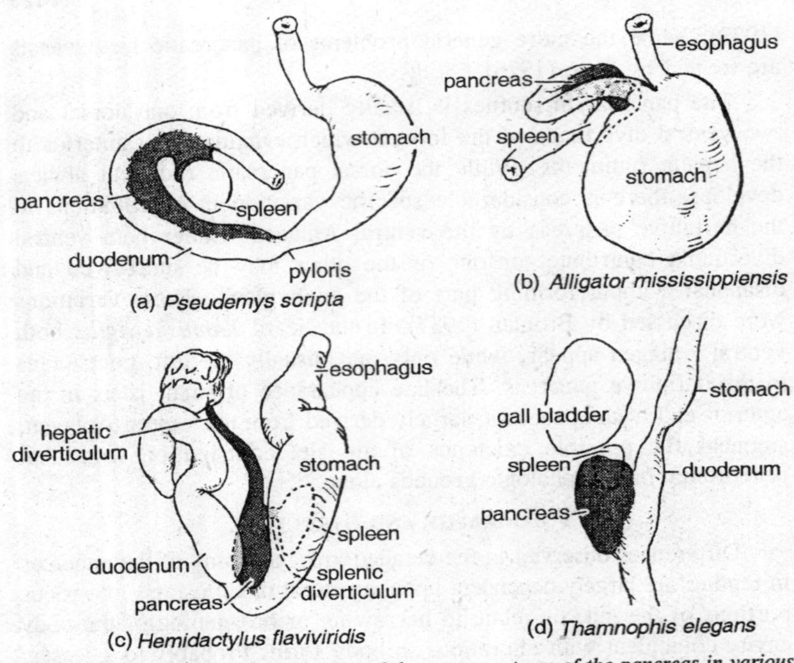

Fig. 6.1. Diagrammatic representation of the gross anatomy of the pancreas in various living reptiles.

artery which originates from the common left-gastric-pancreatic trunk arising from the left aorta. Branches of this vessel supply the pancreas, duodenum, gall bladder, cystic duct, and also a portion of the right hepatic lobe. The spleen and the portion of pancreas applied to its capsule are independently vascularized by a splenic artery, also arising from the left aorta.

We were able to examine only a single living pleurodire, *Podocnemis unifilis*, in addition to poorly preserved museum specimens of *Podocnemis* sp., *Chelodina longicollis*, *Chelus fimbriatus* and *Emydura latisternum*. The arrangement in *P. unifilis* is remarkably similar to that described for the cryptodiran turtles: a wide duodenal loop has the pancreas lying on the mesenteric surface, supplied by a pancreatic-duodenal artery derived from the left aorta. However, the lack of association between the pancreas and spleen is a prominent difference. The latter organ lies dorsal to the gastric curvature, with a left-sided retroperitoneal attachment, and is vascularized by a left gastrosplenic artery supplying the spleen and lesser curvature of stomach. Other pleurodiran turtles examined were museum specimens in all of which

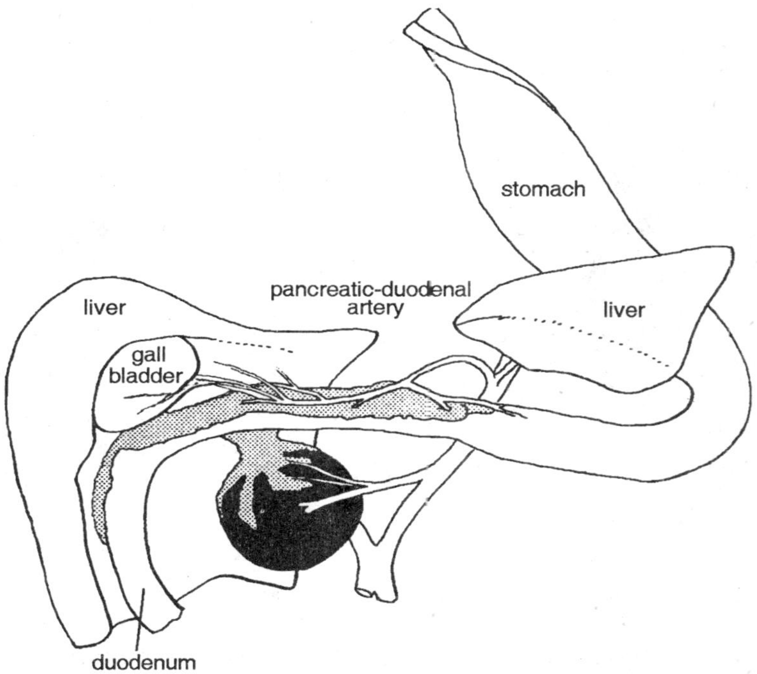

Fig. 6.2. Lissemys punctata, a cryptodiran turtle, showing the spleno-pancreatic relationships and vascularization in ventral view.

the spleen was located on the left side of the peritoneal cavity in close association with the gastric curvature. In *C. longicollis*, the pancreas splayed out over the spleen, but in its most proximal portion, at the level of the pyloro-duodenal junction; in contrast *P. scripta* has the most distal portion of the pancreas in close contact with the spleen.

These preliminary observations suggest the utility of a more thorough comparison of the anatomic arrangements of pancreas and spleen in cryptodiran and pleurodiran turtles. No comprehensive statement on these two groups should be attempted on the basis of the present limited material.

Crocodilians

In *Alligator mississippiensis*, the stomach is large and spherical. The duodenum arises from the crop-like stomach at a point adjacent, but slightly anterior and ventral to the esophago-gastric junction. To the right of the stomach, the duodenum loops first ventrally and then dorsally. The ventral portion of the pancreas lies between the limbs of

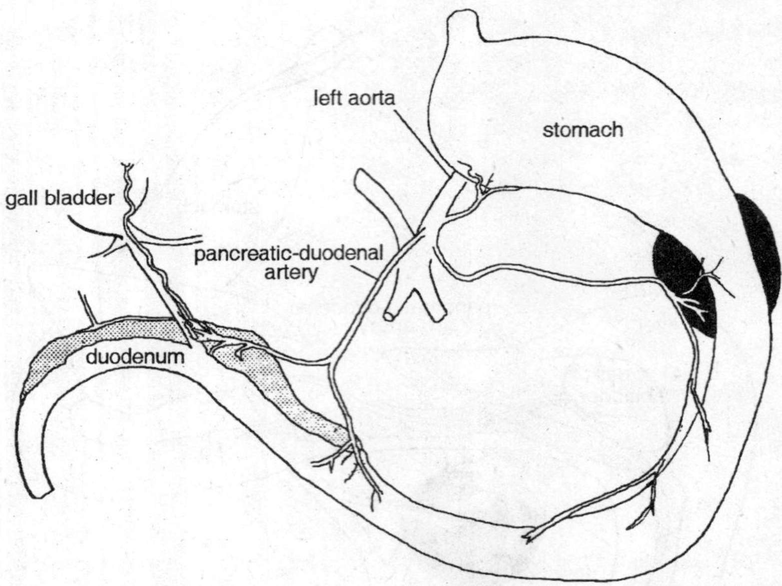

Fig. 6.3. Podocnemis unifilis, a pleurodiran turtle, showing the non associated pancreas and spleen in ventral view.

the ventral duodenal loop, binding them together. The dorsal duodenal loop is formed by the sharp dorsad- and caudad-turning ascending limb. Following the dorsally turning duodenum, the body of the pancreas finally terminates upon the spleen, which is attached to the descending limb of the dorsal duodenal loop.

Lizards

Lizards generally have a pancreas with three limbs, one running along the bile duct towards the gall-bladder, one running to the small intestine, and a usually slender limb, often terminating in a distinct lobe, which runs back to the spleen. In *Hemidactylus*, the duodenum originates at the posterior end of the stomach, curves sharply to the right, and then turns anteriorly to form an ascending loop. A ventral portion of the pancreas lies between the stomach and the ascending duodenal loop, and, stretching dorsally from this portion, a thin extension of pancreatic tissue (splenic portion) ends near or upon the spleen. Extending anteriorly from the ventral pancreas is a thin diverticulum of pancreatic tissue (hepatic portion) encircling the common bile duct. This diverticulum may reach or actually enter the hepatic parenchyma, a feature commonly seen in teleost fish.

Snakes

In snakes somewhat greater attention has been given to the gross anatomy of the pancreas, but no detailed descriptions have yet been published. The relative position of the pancreas in numerous snakes was mapped by Bergman (1961 and earlier), but his reports provide no other anatomical details. Underwood (1967) reported some highly significant and interesting positions of the snake pancreas, but did not deal with the organ's histology.

Because of the potential significance of anatomic variations in the pancreas of snakes, we examined several different types of snakes and were largely able to confirm Underwood's (1967) findings. In many cases, only fixed museum specimens were available for study and their abdominal viscera were sometimes decomposed due to inadequate fixation. This suggests caution in interpreting splenic-pancreatic relationships from such museum material. It should also be stressed that the location, exact form, and anatomic relationships of the pancreas vary markedly even within a single species. Hence studies of the gross anatomy of the ophidian pancreas, should be based upon sizeable samples.

The pancreas of snakes is generally a well consolidated, often pyramidal mass of tissue affixed to the first portion of the duodenum. The pancreas lies posterior to the spleen, though they may be in contact. The pancreas may be just posterior to, or at the same level as, the gall bladder; almost always it lies somewhat posterior to the caudal tip of the liver. Underwood (1967) recognized several structural patterns in snakes which he summarized as: *Cylindrophis*, *Platyplectrurus* and a number of Boidae have a limb of the pancreas running forwards to the spleen. In *Xenopeltis* this isthmus of the pancreas is interrupted. In some species of *Eryx*, in *Sanzinia*, and in *Corallus caninus* there is a distinct gap between the body of the pancreas and the spleen while a splenic limb is missing. *Leptotyphlops* has the spleen applied to the body of the pancreas which has a folded splenic limb. A few lower snakes have the spleen closely applied to the compact pancreas (*Typhlops*, *Boa*, Acrochordidae); all Caenophidia show this condition.

Some of our observations on gross structure suggest modifications. The pancreas of *Typhlops punctatus* was pear-shaped, with the larger portion extending caudally to the right while the spleen lay several millimeters anterior to it, near the anterior end of the gall bladder. The pancreas of *Typhlops proximus* was a spherical, compact body attached to the duodenum but separate from the spleen.

In the single specimen of *Cylindrophis rufus* the pancreas was elongated and divided into three portions. We could not ascertain whether or not there was an interconnecting piece of pancreatic tissue. In *Xenopeltis unicolor*, the pancreas was a solid, pyramidal structure lying against the cranial surface of the duodenum and separate from the spleen.

In the boid *Lichanura roseofusca*, the pancreas was elongate, extending anteriorly from the duodenum between the gall bladder and the stomach, and terminating upon the spleen. In some specimens of *Charina bottae* the pancreas was a single solid mass, but in others it may be divided into distinct portions. It is always in contact with the spleen. The pancreas of *Calabaria reinhardtii* is elongate, extends anteriorly from its duodenal attachment, and lies between the stomach and gall bladder. The anterior end is not in contact with the spleen. The pancreas of *Eryx johnii* was a solid pyramidal body lying ventral in close contract with the duodenum. The spleen was not identified in our specimens.

As noted by Underwood (1967), the pancreata of typhlopids, leptotyphlopids, boids and anillids often appeared to be separated from the spleen. In addition, it appears that in these primitive snakes the pancreas tends to be more elongate and less compact than in boids and Caenophidia. We also confirm Underwood's (1967) report that in the species of Caenophidia the pancreas was usually a compact pyramidal body fixed to the duodenum and in contact with the spleen. Yet in the sea snake, *Lapemis hardwickii*, the pancreas was elongated and lay between the gall bladder and stomach. Its spleen was most unusual in that it was also elongate and wrapped about the entire right edge of the pancreas.

Amphisbaenians

The gross structure of pancreata has been noted in only a few species of this problematic group. In the amphisbaenid *Amphisbaena caeca*, the pancreas is a compact, pyramidal organ resting on the cranial surface of the duodenum, closely adjacent to the spleen. In the trogonophid *Diplometopon zarudnyi*, a compact truncate part of the pancreas is attached to the duodenum, but in addition irregularly-shaped extensions of the pancreas are loosely wrapped about the duodenum.

Thus while the gross pancreatic anatomy of *Amphisbaena caeca* reminds one of that of higher snakes, the arrangement seen in *Diplometopon zarudnyi* is unusual. Definitive statements concerning the amphisbaenian pancreas must await the study of more species and specimens.

HISTOLOGY

Exocrine Pancreas

The exocrine parenchyma of the reptilian pancreas consists of branching tubules rather than the typical acini observed in mammals. Light and electron microscopic examinations of longitudinally sectioned glands revealed parallel rows of polarized zymogen cells facing a small, uniform lumen, while transversely sectioned glands exhibited a histology not unlike that of a true acinus.

First-order exocrine ducts merged abruptly with the tubular parenchyma. In transverse sections, epithelial cells of the duct simply replaced zymogen cells in the parenchymal tubule for a short distance. There is no intercalated or transitional segment or centro-acinar cell as in the mammalian pancreas.

The cytology of zymogen cells in reptiles is similar to that described for all other vertebrates. Zymogen cells are arranged in a polarized position about the glandular lumina: there is an apical portion which is usually packed with eosinophilic zymogen granules, and a basal portion containing the spherical nucleus with a prominent nucleolus. The basal portion of the cell has an intense affinity for basic dyes such as hematoxylin, and lies adjacent to a capillary surface. Ultrastructurally the basophilia is associated with uniform, closely stacked cisternae of the rough-surfaced or ribosome-studded endoplasmic reticulum, known to be the protein-synthesizing intracellular organelle. The apical cytoplasm is often nearly filled with uniform oval profiles of electron-dense, membrane-bound zymogen granules and less electron-dense, somewhat larger precursors or protozymogen granules. The latter granules are formed by condensation of the protein content of the rough-surfaced endoplasmic reticulum in several specialized clusters of cisternae (the Golgi apparatus) with which they are connected. The mature zymogen granule is discharged by fusion of the membrane envelope of the granule with that of the cell surface (emiocytosis). The luminal border of the zymogen cell contains microvilli and stereocilia; adjacent cells are bound by junctional complexes.

The process of protein secretion, granule formation, and granule discharge is basically similar in zymogen cells of all vertebrates, as well as in other protein-producing endocrine cells, e.g., those of the islets of Langerhans, the adenohypophysis, and the juxtaglomerular granular epithelioid cells. Accordingly there were few significant differences in the histologic organization or ultrastructural cytology of the exocrine pancreas in the reptiles examined from that described for the Mammalia.

Microscopic examination of poorly nourished animals characterized by lean appearance, atrophy of the dorsal musculature, and resorption of fat bodies, reveals thin, irregular, dilated exocrine glands. Zymogen cells are atrophic, without apical granulation or the characteristic basal basophilia, and contain numerous cytoplasmic vacuoles and dense basophilic bodies. At the ultrastructural level, numerous large autophagosomes (cytolysosomes) incorporate portions of parallel cisternae of the rough-surfaced endoplasmic reticulum (RER), zymogen granules, and mitochondria. Some zymogen cells were almost completely replaced by these membrane-bound autophagosomes and residual bodies. Cannibalization of cell organelles of the digestive organs of starving animals may represent a method of utilizing endogenous protein. Focal areas of degenerating exocrine glands were infiltrated by foamy phagocytes and inflammatory cells. Thomas (1942) reported similar degenerative changes in zymogen cells of numerous North American snakes, and comparable light and electron microscopic pathology has been observed in experimental pancreatic atrophy induced by protein-deficient diets in mammals and in severe protein malnutrition in infants or kwashiorkor.

Islet Tissue

Reptilian islet tissue, unlike that of most mammals, lacks a sharp demarcation from the exocrine pancreatic tissue. It is typically associated with first-order exocrine ducts and tubules, and lacks a capsule. Ophidian islet tissue, in particular that of more advanced snakes such as the Colubridae, is essentially restricted to the splenic pole of the pancreas, where semi-confluent giant islets are associated with numerous first-order exocrine ducts. Many of these ducts seem to end blindly in islet tissue. Not only are isolated endocrine islet cells intercalated in the epithelium of pancreatic ducts, but better organized, larger masses of islet tissue also lack fibrous or basement membranes separating them from the exocrine tubules. Titlbach (1966) noted a similar dual distribution in turtles. Tight junctions are often observed between exocrine and endocrine cells at the ultrastructural level.

Two general histologic patterns are apparent in reptilian islet tissue. In the first, seen in the Squamata, islet tissue is localized primarily in the splenic portion of the pancreas. Large, irregularly branching cords of islet tissue characteristically occur just beneath the pancreatic capsule, adjacent to the spleen, and, occasionally in some snakes, within the spleen itself. The cords are composed of tall columnar cells with pronounced perivascular polarization, basal nuclei, and apical secretory granules. In contrast to islet organization in mammals, and in some birds and fish, no segregation of alpha and beta cells exists in the Squamata: these two cell types lie adjacent to each other in an alternating pattern along vascular spaces. The amphisbaenians examined in this study (*Amphisbaena caeca*, *Diplometopon zarudnyi*, *Bipes biporus*) did not deviate significantly from this pattern. In snakes belonging to the Boidae and Typhlopidae (*Charina bottae*, *Lichanura roseofusca*, *Typhlops* sp.), however, the islet tissue is more compact with fewer branching cords extending into the exocrine pancreas. Beta cells of *Lichanura roseofusca*, appear to be restricted to the periphery of globose islets along perivascular borders, while the more avascular central portion is formed almost exclusively of alpha cells. A similar arrangement, occurs in *Vipera*.

In contrast to the specialized morphology of the Squamata, representatives of the more "conservative" reptilian orders, the crocodilians and turtles, exhibit a marked segregation of alpha and beta cells, reminiscent of similar degrees of segregation observed in fish, in some birds and in mammals.

The islet tissue of *Alligator mississippiensis* is arranged in loose, globose masses, frequently associated with first-order exocrine ducts at the periphery. Beta cells are often aggregated in the central portion of the islet, while alpha cells cluster peripherally. Segregation is not as pronounced as in some rodents, but is as well developed as in many other mammals, including man. Cytologically, the endocrine cells are more cuboidal than those of Squamata and lack their strong perivascular polarization. Chelonian islet tissue is similarly segregated to a greater (*Geoemyda pulcherrimma*) or a lesser (*Pseudemys scripta*) degree, and there is no concentration of islet tissue in the splenic pole of the pancreas. Titlbach (1966) described peripheral localization of alpha cells in *Emys orbicularis* and *Testudo graeca* as similar to that in *Alligator*.

Only two types of cells with known endocrine functions occur in islet tissue. These are characterized by specific histochemical and immunofluorescent staining reactions, specific ultrastructural configuration of secretory granules and organelles, and experimental identification based on response to cellular toxins or physiologically altered states. The three di-cystine linkages in the insulin molecule form the basis for a number of histochemical procedures, some relying on oxidation of the cystine, which then reacts with the stain. The latter techniques include performic acid-alcian blue, and similar methods for the demonstration of SS/SH groups. Aldehyde fuchsin and chrome alum hematoxylin behave alike after oxidation, but the histochemical basis for these reactions is less well known. In addition, metachromasia produced by the fluorescent dye pseudoisocyanin is used empirically to tag insulin-containing secretory granules in beta cells. Pseudoisocyanin requires oxidized cystine bonds, thus resembling reagents used for identifying SS/SH groups. Alpha cells secrete glucagon in birds and mammals, and apparently in reptiles, and their massed secretory granules impart in acidophilia which is demonstrable with acid dyes (eosin, phloxine, ponceau d'xylidine). Glucagon has recently been identified in pancreatic extracts from turtles in amounts similar to those found in mammalian extracts, and is apparently immunochemically related to the mammalian hormone. The alpha secretory granule also possesses a variable affinity for metallic silver and phosphotungstic acid after oxidation. This argyrophilia of the alpha granule has given rise to claims that two types of alpha cells may be differentiated in many vertebrate pancreata, A2 alpha cells are non-agyrophilic and in many species represent the predominant type of alpha cell. Recently several investigators concluded that the agyrophilic alpha cell (or A1 cell) is identical to the D cell of Bloom (1931), while others have demonstrated exactly the reverse situation in snakes and lizards: the major alpha cell is agyrophilic and a D cell is nonagyrophilic. We have confirmed the agyrophilia of alpha cell in *Leiocephalus* using the method of Lee (1967), who similarly reported rabbit alpha cells to be agyrophilic. In the chimaeroid fish *Hydrolagus*, all but 5% of the non-beta-cell islet elements, alpha, D and X, can be agyrophilic, depending on the silver staining method used. Despite arguments supporting the unitary nature and constant agyrophilia of the Al/D cell, the body of evidence suggests that the ability to deposit silver in proteinaceous secretory granules is probably not dependent upon any basic structural similarity of the molecule. These capricious techniques differ significantly in kind from those which rely on the known di-cystine links in insulin to characterize beta cells. Honma and Tamura (1968) discuss the non-specificity of the silver impregnation methods.

D cells comprise a proportionately small component of islet tissue. They do not react to the usual staining methods for beta cells nor to acid dyes, but will stain with Light Green. They have a variable affinity for silver impregnation. Gomori (1941) noted the distinction between these cells and typical alpha cells, as well as the preferential localization of both cell types together in the acini of human pancreas. He suggested that the D cell was an altered alpha cell, and ultrastructural studies have supported this early observation.

Some recent investigators have suggested on the basis of reported tinctorial similarities to a gastrin-producing human islet cell neoplasm, that the delta cells secrete a third pancreatic hormone. However, ultrastructural studies of these tumors do not support this viewpoint. Certainly some human endocrine neoplasms retain the ability to synthesize the original hormone. However it should be noted that others, particularly bronchogenic carcinoma and islet cell tumors, can produce *de novo* a host of hormones normally synthesized elsewhere; these include ACTH and parathormone, gastrin, and, as recently suggested, secretin in the case of islet cell tumors. Yet there is no physiologic evidence that the human pancreas normally produces ACTH, parathormone, gastrin, or secretin. Neoplastic hormone production should not be strictly interpreted to imply that the normal parent tissue synthesizes that hormone.

Very recently Lomsky et al. (1969) demonstrated by an indirect fluorescent antibody technique that mammalian islet cells, identifiable as agyrophilic D or A, cells, correspond to cells which bind *rabbit anti-porcine gastrin*. This preliminary work, as noted by the authors, contradicts earlier attempts to isolate gastrin from a number of normal mammalian pancreata. In addition, McGuigan (1968), using a similar fluorescent antibody technique to identify gastrin-containing cells in hog antrum, was unable to demonstrate agyrophilia in these cells. The singular finding of cells ultrastructurally consistent with enterochromaffin type II cells in rabbit islets lends substance to the theory that rare heterotopic gut elements can be present in islet tissue. To what extent these heterotopic elements exist, and to what degree they correspond to *gastrin* producing and/or agyrophilic D or alpha cells remains to be seen.

Considerable variation occurs in the amount of stainable granules in representatives of any one cell-type, and some cells appear "clear" or devoid of granules. Recent studies tend to subdivide alpha and beta cells according to the degree of granulation, usually on the basis of the intensity of the staining reaction. While such subdivision is useful in characterizing types in terms of the animal's physiologic state, the misinterpretation that different types secrete different protein hormones should be avoided.

To help clarify the morphologic classification of reptilian islet cell types, we have studied the fine structure of the pancreatic islets of *Leiocephalus carinatus*, *Contia tenuis*, *Drymarchon corais*, *Elaphe vulpina*, *Alligator mississippiensis*, and *Pseudemys scripta*. Fixation for electron microscopy was achieved by canulation and perfusion of the left aorta in anaesthetized animals. In turtles the pancreatico-duodenal arterial trunk was used. The perfused fixative was 1.5% distilled glutaraldehyde in 0.067M sodium cacodylate buffer and 1 gm % sucrose at pH 7.4, 310 mOsm and 32-35°C. Tissue became fixed within a minute of perfusion.

Osmium tetroxide was used as a post fixative. Araldite imbedded tissue was sectioned at 800 Å, sections were stained with uranyl acetate and lead citrate and viewed with a Siemens Elmiskop I electron microscope. Araldite sections for light microscopy were sectioned at 0-5 μ and stained with toluidine blue. Only the general features of reptilian pancreatic islet ultrastructure derived from these observations are considered here, in relation to a general discussion of the fine structure of this tissue in reptiles.

The ultrastructural morphology of secretory granules best distinguishes the cell-types of endocrine islet tissue. Beta secretion granules are unit-membrane-bound and characteristically contain polyhedral crystalloid condensations within a lighter matrix. Such condensations are peculiar to the beta secretory granule and permit identification when the size, density and unit-membrane profile of alpha granules are similar. The condensations present tetrahedral, trapezoidal, hexagonal, and irregularly polyhedral profiles. Greater magnification frequently discloses a substructure of stacked, parallel, linear densities or plates, clearly shown in *Drymarchon* in which a constant periodicity of 62 Å was measured. This measurement compares with plate periodicities of 36 and 18 Å in the salamanders *Amphiuma* and *Ambystoma*, respectively, and 15 Å in the dog. Differences in the periodicities in the crystalloidal substructure of granules in beta cells may reflect differences in the molecular size of the insulin. In a similar problem with another protein, Lessin (1968) found a close correspondence between the measured periodicity of crystalloidal aggregates of hemoglobin in red cells obtained by freeze-etching electron-microscopy and those obtained by X-ray diffraction, confirming that individual plates correspond to single bimolecular layers. He notes that the usual fixation and embedding procedures cause considerable artefactual shrinkage of the plate width, apparently due to dehydration and polymerization. Further, the abnormal human hemoglobins, types S and C, which differ by only one amino acid in composition from the normal type A, produce a small but reproducible difference in their crystalloidal periodicity. The difference observed in the crystalloidal substructure of reptilian beta granules may, indeed, reflect differences in the amino acid sequence of reptilian insulins, as well as considerable but at present unassessed artefact. But aside from fish differences in amino acid sequence are small, particularly in mammals; while chicken insulin differs from that of the Sei whale by only two amino acid residues in the A and three in the B chain. Unfortunately the molecular size, weight, amino acid composition, and sequence is not yet known for any reptilian insulin, although an educated, but incautious guess would make them similar to the insulins of chickens and other birds. In the snake, *Drymarchon*, the periodicity of beta plates does not vary with the size, shape, or density of the secretory granule in which it occurred.

Other, presumably less condensed or immature, beta secretory granules have large spherical profiles and a homogeneous flocculant content. In *Drymarchon*, the less condensed granules exhibit a complete series of transitional forms toward the dense crystalloid granule. These forms develop areas of marginal electron-density, which in other profiles reveal circumferential confluence of dense areas with a residual, less dense, central matrix. The polyhedral aspect of the secretory granule may appear before any peripheral condensation has occurred, implying that a crystalloid structure is present even before condensation begins. In beta cells with strong perivascular polarization, no preferential stratification of one or the other type of granule was noted. We found multiple, discrete, polyhedral condensations in beta secretion granules of *Alligator* and *Pseudemys*. A similar morphology was evident in *Lacerta*, while in certain Squamata (*Drymarchon*, *Elaphe*, *Contia* and *Leiocephalus*) beta secretory granules contain a single crystalloid condensation.

In contrast to beta secretory granules, the granules of alpha cells are indistinguishable from many proteinaceous secretory granules found in non-islet secretory cells. Alpha granules characteristically have round profiles; they are usually homogeneous, but occasionally reveal a paracrystalline substructure and nearly fill the unit membrane. A thick zone of lesser electron-density lies between the granule and the unit membrane. Although the condensed portion of the beta secretory granule is often smaller than the alpha granule, mean diameters of the profiles of alpha and beta unit membranes are almost identical in all species examined except *Drymarchon*, in which profiles of alpha unit membranes are twice the diameter of beta profiles. Elongate spheroids and rod-like profiles of alpha granules also occur.

Almost every alpha cell contains unit-membrane-bound granules of low electron-density, without central dense cores. In fact many contain only a slight electron-dense peripheral condensation on the inner surface of the unit membrane, similar to the peripheral condensation zone of typical alpha granules. A central portion of the granule consists of loose fibrillar material indistinguishable from that of typical D cell granules reported by Sato et al. (1966) and Fujita (1968). A continuous gradation was observed between the condensed alpha granule and the loose fibrillar form within single cells. In addition, individual alpha cells vary considerably in the proportion of loose, fibrillar or dense-type granules. Some cells contain only one or two typical dense alpha granules, while the remaining ones are of the loose, fibrillar D-type. Finally, some cells, indistinguishable from previously described D-cells in reptiles, contain only the loose, fibrillar-type granule. The genera with D cell type of morphology studied by us (*Leiocephalus*, *Contia* and *Alligator*), always showed a complete range of intergrades between typical dense alpha, and loose, D cell granular profiles. Cells with a granular morphology typical of D cells, commonly occur in association with alpha cell clusters, as noted previously by Gomori (1941), Sato et al. (1966) and Honma and Tamura (1968). Conversely, beta cell clusters rarely reveal a morphology characteristic of D cells. In *Leiocephalus*, the frequency of profiles of the loose fibrillar D type granule in alpha cells is paralleled by the frequency of autophagosomes and complexes of membrane-bound lipids and myelin figures. Some alpha D cell intergrades are filled with autophagosomes and contain few granules. Since these cells border well-perfused capillaries and lie adjacent to well-fixed alpha cells, they are probably not artefacts.

Unquestionably, granules with the D type of ultrastructure can occur and presumably can be synthesized in otherwise unmistakable alpha cells. We have demonstrated a complete gradation between alpha cells and D cells in some reptiles, like that previously reported in man. However, Machine and Sakuma (1967) and Machino et al. (1966) have presented clear evidence of active emiocytosis, discharge of secretory product, in D type granules of chick embryos, thus suggesting a functional cell, and Przybylski (1967) found similar granular profiles in differentiating alpha cells in the chick embryo. The fortuitous occurrence of autophagosomes and D type cells in *Leiocephalus* may suggest a common functional condition in such cells which is not shared by typical alpha cells. This association should not be misinterpreted as a general indication of a moribund state; the majority of such cells in other species we studied lacked such lysosomal activity.

A recent paper by Kobayashi and Fujita (1969) which appeared during revision of this manuscript confirms our own impressions of the artefactual nature of most descriptions of D-cell ultrastructure. These authors also employed perfusion fixation a D and A cell side by side we interpret as the type of intergrade observed in our own studies. We would interpret both as A cells.

In *Alligator*, columnar cells bordering the pericapillary spaces but lacking specific secretory granules resemble in their ultrastructure the C cells described by Munger et al. (1965) and Sato et al. (1966). In the former, however, these cells possess a villous apical surface partially forming the wall of a small first-order exocrine duct. Adjacent zymogen cells display ciliated surfaces, and zymogen granules are emptied into the duct, which is filled with granular electron-dense material. In *Contia* numerous additional examples of the epithelium of first-order ducts are intercalated between islet cells bordering a pericapillary space. Although some mammalian C-cells may be degranulated or quiescent endocrine cells, and despite attempts to equate all C-cells with granular D-cells, reptilian C-cells may form the epithelium of first-order duct, even though the luminal border does not appear in all sections examined.

Aside from secretory granules, some islet cells contain certain other distinguishing organelles. Mitochondrial profiles in beta cells are curved and tubular, with irregularly disposed, transverse cristae and granular, electron-dense matrices. Alpha cells contain similar mitochondria, but the snake *Contia* possesses numerous large, relatively rigid mitochondrial profiles with longitudinally-oriented, parallel cristae. A few sections reveal abrupt transitions between the longitudinal and transverse cristae within a single mitochondrion. At higher magnification, the cristae are seen to bear a beaded fuzz of rigid periodicity; in transverse section, the cristae themselves have triangular profiles with coronae of small beaded densities. The repeating fuzz on these specialized reptilian mitochondria resembles the repeating particle associated with the inner membrane of mitochondrial cristae reported by Fernandez-Moran et al. (1964). Similar prismatic cristae in the mitochondria of specialized reptilian alpha cells were described by Burton and Vensel (1966), but mitochondria with similar rigidly parallel, longitudinal cristae have been reported in the beta cells of toads and mammals, in neurons of fish, and in the cricothyroid muscle of bats. The periodicity of intracristal and intercristal mitochondrial suborganelles was reviewed by Suzuki and Mostofi (1967). In *Contia*, there is a filamentous, paranuclear macular zone similar to that described in the beta cells of toads and mammals and mammalian alpha cells. We saw no strict association of specialized mitochondria with the macular zone in *Contia*.

Bencosme (1959) first described secretomotor innervation of islet tissue, and recent studies have revealed both adrenergic and cholinergic secretomotor innervation of alpha and beta cells in some mammals and in toads. Physiologically, corroboration of this innervation is not yet complete. We ourselves observed secretomotor innervation only in snakes (*Drymarchon* and *Contia*), and then only in beta cells. The secretomotor synapses often lie depressed in the surface of the beta cell, beneath the parenchymal basement membrane. Clear synaptic vesicles measuring ca. 500 Å occur in clusters, while there are relatively few large, dense-cored vesicles 1200-1500 Å in diameter. Such morphology suggests that the synaptic ending is *adrenergic*. In addition to secretomotor synapses, multiple unmyelinated nerve profiles were observed in the pericapillary spaces of all species. Fenestrated endothelium of the islet capillaries and their perivascular connective tissue spaces are characteristic of endocrine organs.

Conclusion

In this brief review of the reptilian pancreas we attempt to collate the existing literature and our own limited direct observations in a form which reflects current areas of research in this field. We hope the report provides strong support for the theory that any such study benefits from a comparative viewpoint: many of our observations of reptilian pancreatic morphology become relevant only by reference to literature dealing strictly with mammals or for that matter, with fishes.

In numerous instances, we have postulated our own interpretation, particularly in regard to the ultrastructure of islet tissue, where we hope to provide a re-examination of some prevailing notions. Though strongly worded, our interpretations are not intended to masquerade as ultimate or definitive statements, but rather to stimulate further investigation in this area.

7

Thymus Gland

The thymus is an organ unique to vertebrates. Correlated with its appearance is adaptive immunity—the ability (1) to react specifically against foreign material by cellular proliferation and antibody production and (2) to retain a "memory" of that reaction which will result in heightened response upon being challenged with the same foreign material again. When the thymus is ablated during a critical period of the development of an individual, adaptive immunity will be but partially expressed. Therefore, the thymus seems necessary for proper development of the immune response.

Reptiles, like most vertebrates, show adaptive immunity. Study of the reptilian thymus, however, has not progressed as far as has that of mammals and birds, and the role of the reptilian thymus in development and maintenance of adaptive immunity must, for the present, be based on inference, approached through careful comparison with other vertebrates.

The immunological deficiencies caused by neonatal thymectomy of mice may be corrected by implantation of thymus fragments contained within cell-impermeable diffusion chambers. It seems reasonable, therefore, to consider the thymus a gland capable of producing some humoral factor important for immunological maturation.

Several excellent reviews summarize the older investigations of reptilian thymus. The present discussion is not meant to recapitulate these. Rather, I present generalizations on the morphology typical of the various reptilian groups and point out deviations from that pattern. Light microscopic anatomy is correlated with recent ultrastructural observations of the reptilian thymus as well as with the fine structure of the thymus in other vertebrates. It is hoped that this approach will establish a reasonable basis for subsequent extension of knowledge of the thymus, both in reptiles and in other animals.

General Morphology

Reptilian thymus glands typically are bilateral organs which occupy a variable area of the neck in close relationship with the large vessels and nerves. The thymus of each side frequently consists of two small white or yellowish lobes which are completely separate from each other. These lobes may be termed anterior and posterior according to their disposition along the longitudinal axis of the body, or they may be given the number of the pharyngeal pouch from which they are derived. There are differences in the number and shape of thymic lobes within the major reptilian groups, but a characteristic anatomical situation may be described for each group.

Lepidosauria

A "typical" thymic morphology for the tuatara and lizards may be derived from studies of *Sphenodon punctatus*, *Lacerta* sp., *Chalcides ocellatus*, and *Scincus scincus*.

Two apparently non-lobulated lobes, a cephalic thymus 2 and a caudal thymus 3, lie lateral to the pharyngeal wall on each side. The lobes may lie within richly pigmented connective tissue chambers which are distinct from the thymic capsule proper. Thymus 3 is larger and reaches the carotid arch posteriorly. Thymus 2 is situated close to the base of the skull, dorsal to the laterally projecting recessus piriformis,

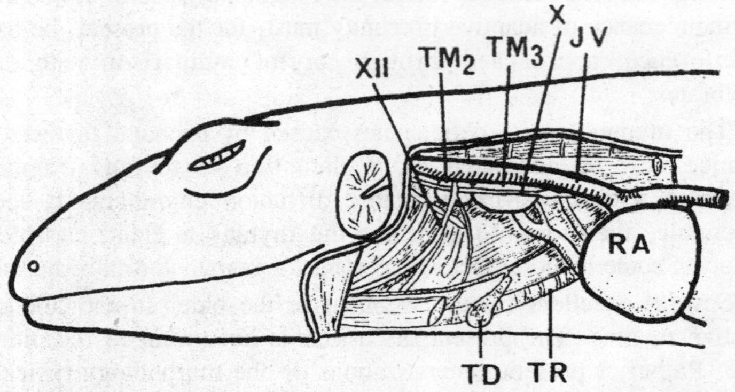

Fig. 7.1. Location of the thymus in the lizard, Lacerta agilis. The two thymic lobes (TM2 and TM3) lie in close relationship to the vagus nerve (X) and jugular vein (JV). The anterior lobe (TM2) is crossed by the hypoglossal nerve (XII). TD, thyroid; TR, trachea; RA, right atrium.

and thus is not evident in ventral view. The hypoglossal nerve crosses thymus 2 superficially.

Both lobes lie ventral to the internal carotid artery and medial to the internal jugular vein and vagus nerve; the nodose ganglion of the latter is located close to the posterior margin of thymus 3.

The relative size of the two lobes varies between individuals as well as between species. The lobes may overlap each other slightly at their opposed extremities or be separated by a considerable distance. They may be elongate and flattened, oval, or cone-shaped. Except for such minor variations, however, a "typical" saurian thymus has been described, in addition to the examples cited above, in *Iguana iguana*, *Ophisaurus aspodus*, and *Gekko gecko*; *Uromastyx*, *Agama*, and *Acanthodactylus*; and *Varanus*.

Greater variation may occur in specific cases. Some specimens of *Uromastyx*, *Agama*, and *Acanthodactylus* have only one lobe on each side. The thymus of *Psammodromus* consists of six or seven small lobes which are distributed over almost the total length of the neck; the lobes are completely isolated except for a tract connecting the units with one another. The thymus of *Psammodromus* is thus more similar to that of birds than to that of a "typical" lizard. A Y-shaped thymus occurs in *Amphisbaena*, the unpaired posterior portion presumably resulting from a fusion of right and left sides.

Salkind (1915) described one small, whitish thymic lobe just posterior to the mandible on each side of the neck in *Chamaeleo* sp. However, Pischinger (1937) found only a thin, undivided, roundish strand stretching along the great vessels from the base of the skull to the thoracic aperture and compared his findings with a similar condition described by Simon (1845) for the Lacertidae and Gekkonidae. Since Simon does not state what species he studied, it is impossible to tell whether this represents species differences or some other variation.

The unlobulated thymus of snakes is located immediately anterior to the heart. Two lobes lie between the median thyroid gland and the lateral fat body of each side in *Natrix natrix* and *Coronella austriaca*. The lobes are closely associated with the common carotid artery, jugular vein, and vagus nerve; the nodose ganglion is situated near their posterior ends. The thymus may be asymmetrical, with the lobes of one side larger than those of the other.

The number and disposition of lobes varies as in lizards. Only a single elongate lobe on each side is found in some specimens of *Boa*, *Python*, *Coluber*, *Hydrus*, and *Crotalus*. In *Malpolon*, the thymus is

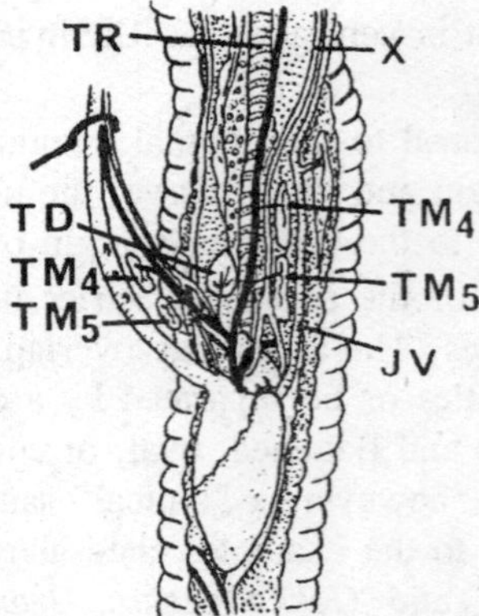

Fig. 7.2. Location of the thymus in the grass snake, Natrix natrix. The two thymic lobes (TM4 and TM5) of each side lie close to the vagus nerve (X) and jugular vein (JV) immediately anterior to the heart. TD, thyroid; TR, trachea.

composed of confluent lobes that may be enclosed in fat, simulating a single median organ.

Testudines

The thymus of turtles appears lobulated. A fairly constant thymus 3 lies, on each side, in the angle formed by the division of the subclavian and common carotid arteries. Thymus 4, which is always smaller and variable in development, may not be evident. The vagus nerve lies lateral to the thymus lobe. The nodose ganglion, like that of lizards and snakes, is located at the posterior end of the gland.

Except for minor variations in lobe structure, the above description applies to relatively mature examples of *Emys orbicularis* and several

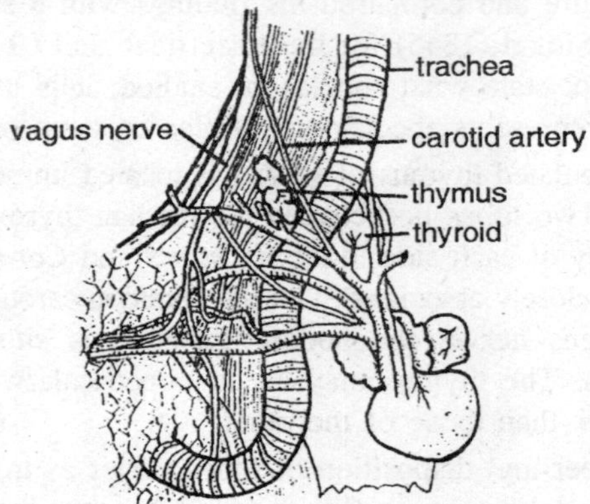

Fig. 7.3. Location of the thymus in the tortoise, Geochelone denticulata.

species of *Testudo*. In young *Chelonia mydas*, however, Van Bemmelen (1888) described the thymus as a larger mass stretched out over the posterior third of the neck, covering the carotid artery, and broadened at its posterior end. This situation is like that normally encountered in adult Crocodilia.

Crocodilia

Crocodiles and alligators have a thymus more like that of birds than that typical of other reptilian groups. Figure 7.4 depicts the elongate left thymus of young *Crocodylus porosus*. A broadened posterior extremity lies immediately anterior to the heart, and a narrower portion traverses the neck region to approach the base of the skull. The common carotid artery is ventromedial and the jugular vein and vagus nerve ventrolateral to the extensive gland on each side of the body.

The reviews by Hammar (1909) and Pischinger (1937) indicate that this basic scheme, with minor variations, is found in all species

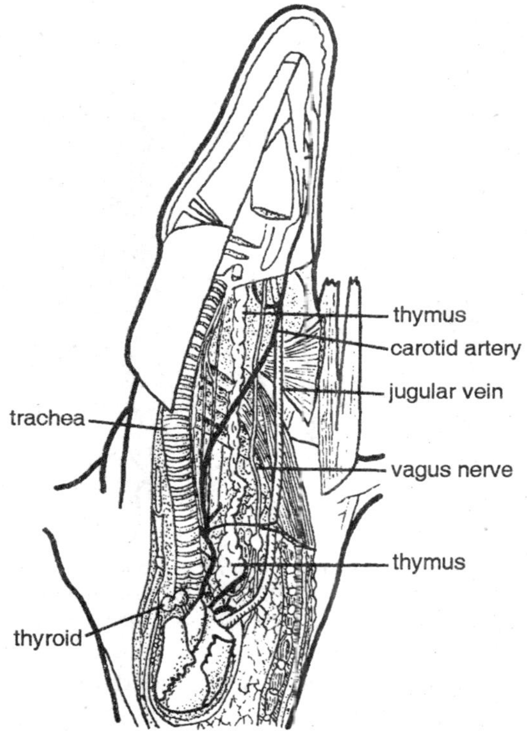

Fig. 7.4. Location of the thymus in a young Crocodylus porosus.

of crocodiles and alligators which have been investigated. Young animals are more apt to have a thymus which occupies the total length of the neck. The prismatic posterior enlargements of the right and left thymus may meet in the midline, covering the thyroid gland. In adult *Caiman crocodilus* and *Caiman latirostris* the anterior end of the thymus may reach only the anterior end of the chest cavity. The latter species may have additional, rounded, cervical lobes.

Embryonic Development

The reptilian thymus originates from the pharyngeal pouches. Like those of fish, Amphibia, and birds, reptilian thymic primordia are dorsal outgrowths of the pharyngeal walls in distinction to the ventral outgrowths which produce the thymus of mammals.

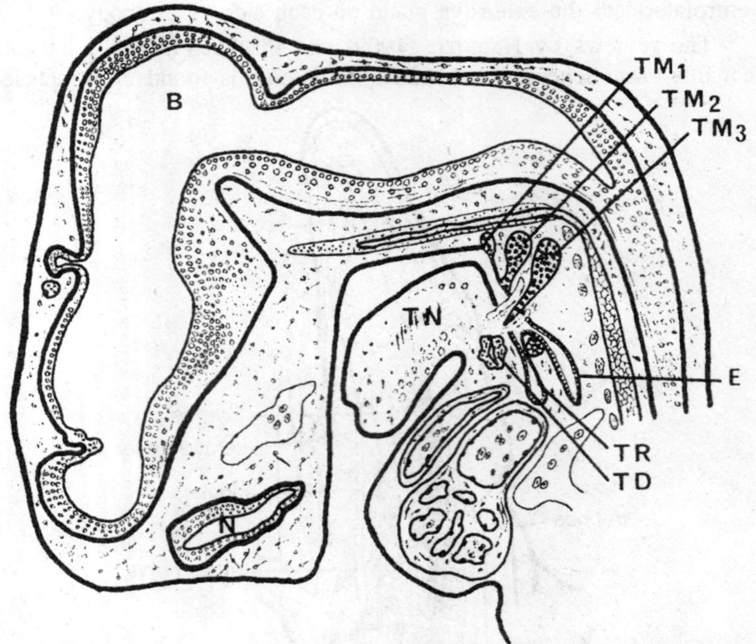

Fig. 7.5. Saggital section of an embryonic lizard (Lacerta agilis) showing dorsal thymic outgrowths (TM1, TM2, and TM3) from the first three pharyngeal pouches. The posterior two are retained in the adult. B, brain; E, exophagus; TD, thyroid; TN, tongue; TR, trachea.

The thymus of lizards, snakes, and turtles develops from different pairs of the five pharyngeal pouches. To my knowledge, there have been no comparable studies of Amphisbaenia or Crocodilia.

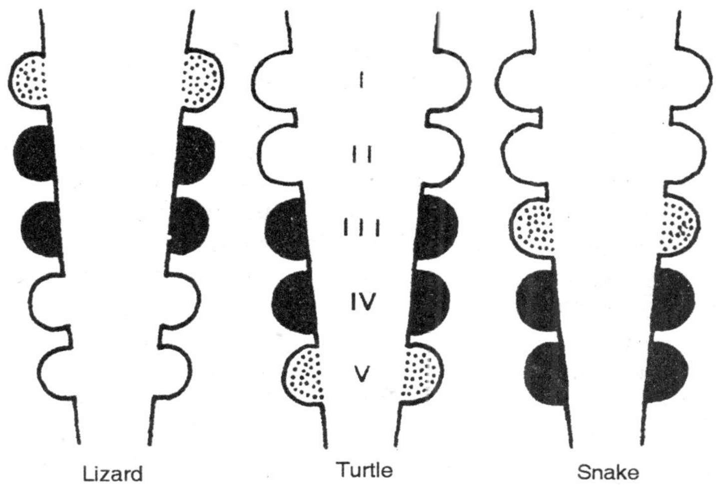

Fig. 7.6. Diagrammatic representation of pharyngeal pouches (I-V) which give rise to transitory (stippled) and definitive adult thymus (solid).

The definitive anterior and posterior thymic lobes in lizards originate as dorsal outgrowths of the second and third pharyngeal pouches. In addition, a transient thymus rudiment associated with the first pouch has been described in *Lacerta agilis* and *L. muralis*. Saint-Remy and Prenant (1904) corroborated that the thymus of *Anguis fragilis*, *Chalcides ocellatus*, *Lacerta viridis*, and *L. agilis* developed from the second and third pouches, but denied the existence of a rudimentary thymus 1 in any of these species. They also noted that some thymic tissue develops from the ventral portion of the third pouch and is incorporated in the posterior lobe (thymus 3).

A constant thymus 3 originates from the corresponding pharyngeal pouch in turtles (*Chelonia mydas*, *Chrysemys picta*). A variable thymus 4 and a transitory bud develop from the fourth and fifth pouches respectively.

A further caudal shift is seen in snakes. The definitive thymic lobes originate from pharyngeal pouches 4 and 5 (*Natrix natrix*, *Elaphe longissima* and *Natrix natrix*). A transitory thymus 3 may appear as a dorsal outgrowth of the third pouch.

Thymic development begins as a slight crescentic invagination of pharyngeal epithelium just medial to the branchial placode. Epithelial proliferation forms a hollow, lobulated, fingerlike projection. Connection with the pharyngeal pouch is broken and the intermediate pouch tissue disappears, so the thymus lies free in surrounding mesenchyme.

Progressive cellular proliferation obliterates the central cavity. The cells take on a typical lymphoid appearance. Thymic lobules become subdivided into distinct cortical and medullary areas, and thymic corpuscles appear in the medulla.

Aberrant collections of thymic tissue may occur at locations distinctly separated from major thymus lobes. Adams (1939), for instance, found a microscopic thymus 4 posterior to thymus 3 (the definitive posterior thymus) in histologic sections of *Lacerta*. Thymic tissue has also been noted within the postbrachial body of the turtle *Pseudemys scripta*.

The thymus and parathyroid glands originate very close to each other in the branchial pouches. It is not surprising that this intimate association is sometimes maintained throughout development. Follicles of flattened, spindle-shaped cells have been described within the thymus of the tortoise and alligator. These follicles were almost certainly the same as the parathyroid with cells "arranged in a whorl" described in the thymus of *Natrix sipedon fasciata* by Thompson (1910). Parathyroid glands are sometimes embedded in the thymus of embryonic and adult turtles and young rattlesnakes.

The cells from which thymic lymphoid cells, or thymocytes, differentiate during organogenesis have not been clearly established. Maurer (1899) concluded from histological evidence that lizard thymocytes derived from endodermal epithelial cells. Experimental studies of developing thymus have recently led Auerbach (1961) to conclude that epithelial cells are the progenitors of thymocytes in mice, and these results were corroborated by electron microscopic investigation of thymic development in the chick. Thus it is likely that reptilian thymocytes arise from epithelial cells but strong supporting evidence is still lacking.

Histology

The thymus of reptiles is surrounded by a well defined capsule of dense connective tissue. Septal extensions from the capsule subdivide the thymus of turtles into distinct lobules. Lepidosaurians and crocodilians characteristically lack such lobulation, although small connective tissue trabeculae are associated with penetrating blood vessels.

Kohnen and Weiss (1964) have pointed out the wide variety of structures which have been described in the literature as thymic corpuscles. The relatively solid, lamellar arrangement of epithelial cells typical of thymic corpuscles in humans is rare or non-existent in reptiles. Rather, cyst-like structures are present. Reptilian thymic cysts range from minute intracellular structures through formations well over 0-1 mm in diameter. Epithelial cells comprise the boundary of the cysts. The central portion of large cysts is frequently occupied by cellular elements in varying stages of degeneration. Granulocytes and myoid cells may be recognized within some. Such formations probably represent foci of cellular degradation segregated from surrounding thymic parenchyma by a continuous layer of epithelial cells.

The predominant cell type, particularly in young animals, is a lymphoid cell, the thymocyte. Large, medium, and small thymocytes are present, the latter in greatest quantity. At the periphery of the thymic lobes or lobules, in the thymic cortex, they tend to be so numerous that they mask the details of other cells. The more central tissue, the medulla, appears less dense because of a reduced concentration of thymocytes and a concommitant higher proportion of other, lighter-staining cells.

Epithelial cells constitute the second principal cell type. They are easily identified in the medulla by their large, relatively clear nuclei, prominent nucleoli, and abundant cytoplasm. Epithelial cells are distributed throughout the cortical area as well, but their cytoplasmic processes are there attenuated between intervening lymphocytes, making identification more difficult. Epithelial cells may occur singly, in aggregates (nests), or in relationship with thymic corpuscles.

Myoid cells are many times the diameter of thymocytes and have long been recognized in the reptilian thymus. They have also been referred to as granular cells, as myoepithelioid cells, as unicellular thymic corpuscles, and perhaps also as enlarged blood corpuscles. Although a concentric patterning is sometimes seen within their cytoplasm, they are clearly not thymic corpuscles, but a distinct type of cell.

Myoid cells are most numerous in the medulla, but do occur in the cortex. They have one, occasionally two, pale-staining nuclei each with one or two prominent nucleoli. The nuclear diameter of a myoid cell usually greatly exceeds the total diameter of a thymocyte. Round or oval cell profiles and elongate forms are found. Cross striations similar to those of skeletal muscle are evident in some myoid cells. The ultrastructural basis for cytoplasmic lamellation and crossbanding will be discussed in another section.

Granulocytes are sometimes prominent in reptilian thymuses. Cells with rounded or slightly lobulated nuclei and eosinophilic granules frequently are numerous in the connective tissue, thymic parenchyma, and cysts. These granules are PAS positive, a condition described for avian eosinophils. Cells with basophilic granules (mast cells) occur in smaller numbers. The latter are found in relation with connective tissue and to a lesser extent in the parenchyma. There is no consensus of opinion as to whether these granulocytes originate within the thymus or penetrate from the blood stream after forming at another site. Jordan and Looper (1928) described, in the thymus of the box turtle (*Terrapene carolina*), cells which contained both basophilic and eosinophilic granules and concluded that cells with basophilic granules develop from thymocytes and that the granules subsequently "ripen into acidophilic conditions". Partial corroboration of this interpretation has been presented recently by Ginsburg and Sachs (1963) who described transformation of mouse thymic cells into mast cells in tissue culture.

Numerous reticular fibers may be shown in the capsule and septa by silver techniques and the periodic acid-Schiff reaction. Such reticular fibers are particularly prominent in association with blood vessels, but may be interposed between elements of the thymic parenchyma as well. They tend to be sparse in areas replete with small thymocytes.

There probably are neither arborizations of nerves nor lymphatic vessels within the thymic parenchyma.

Fine Structure

The fine structure of reptilian thymus is similar in many respects to that described for mammals, embryonic chicks, the axolotl, and frogs. Some of the salient ultrastructural features of the reptilian thymus based on electron microscopic studies of snakes (*Crotalus*, *Lampropeltis*), lizards (*Eumeces*, *Heloderma*), and a turtle (*Trionyx*).

Cell Types

Thymocytes

Most reptilian thymocytes have a diameter of approximately 4 to 6 μ. Although they tend to be spherical, considerable deformation may be produced by surrounding cells. Especially for smaller cells the amount of cytoplasm is small in relation to the area occupied by the nucleus of each thymocyte.

The nucleus may be circular in outline or may mirror the deformation of the cell membrane. The nuclear membrane is frequently indented. Clumps of chromatin are prominent at the periphery of the nucleus, while one or more clumps usually occur centrally, either isolated or continuous with the peripheral chromatin.

The cytoplasm of thymocytes contains an abundance of ribosomes arranged singly or in groups. Rough endoplasmic reticulum is very sparse. A few relatively large mitochondria are usually present. A small Golgi apparatus and centrioles may be seen in some cells.

Epithelial cells

Thymic cells are identified as epithelial primarily by the presence of cytoplasmic tonofilaments and/or desmosomes. The latter two elements are abundant in some cells. In others, they may be so sparse that they are not observed in a particular section. Such cells are considered to be epithelial because of their similarity with other epithelial cells.

One type of thymic cell has a large, electron lucid nucleus which contains one or two prominent nucleoli. Chromatin is relatively evenly distributed throughout the nucleus with some concentration at the periphery. The cytoplasm contains scattered ribosomes, small mitochondria and short elements of rough endoplasmic reticulum. Tonofilaments may appear as isolated bundles in the cytoplasm or in association with desmosomes. A basement membrane occurs at the boundary between epithelial cells and connective tissue.

A much more irregular and dense nucleus characterizes a second type of thymic cell. Coarse, angular blocks of chromatin are scattered throughout its nucleus. The cytoplasm is dense and may contain phagocytic inclusions of varying size, structure, and contents. Slender processes of these cells extend between other elements of the thymic parenchyma. In some sections connections of processes with the cell body are not evident, giving the appearance of isolated fragments of cytoplasm interspersed between thymocytes. Few desmosomes are present.

Small, moderately dense granules approximately 0.2 μ in diameter occupy the cytoplasm of a third cell type whose cytoplasm contains a well-developed Golgi apparatus. The nature of the small granules is unknown. The nucleus is fairly regular in outline and intermediate in density between the two types of epithelial cells described above. There is a morphological similarity between these cells and the beta cells of pancreatic islets. Recent evidence suggests that an insulin-like factor is produced by the "beta" cells of the mouse thymus, and the techniques used to determine this (aldehyde-fuchsin staining, bioassay of thymic extract, immunoassay, immunofluorescence) should, when applied in conjunction with ultrastructural studies, be useful in determining the presence or absence of similar cells or factors in the reptilian thymus.

Another variant of epithelial cell is characterized by a very irregular nucleus with thin, highly elongate nuclear extensions which may exceed 13 μ in length but be only 0.3 μ in diameter. The chromatin is dense and concentrated at the margins of the nuclear extensions. Cytoplasmic organelles are similar to those of other epithelial cells. This cell type tends to occur in groups, with the filamentous nuclear extensions and surrounding cytoplasm running parallel for some distance.

Myoid cells

The fine structure of thymic myoid cells has been studied recently in many species of reptiles, birds, amphibians, and mammals.

Myoid cells in the reptilian thymus may be rounded to highly elongate. Spherical myoid cells are very large, frequently 20 to 30 μ in diameter with a centrally placed nucleus 4 to 8 μ in diameter. Myofibrils are disposed circumferentially around the nucleus in various planes, so that longitudinal, transverse, and oblique profiles are observed in a single section. The myofibrils of elongate cells tend to lie parallel to the longitudinal axis of the cell.

The myofibrils usually are composed of thick and thin myofilaments like those of skeletal muscle. Raviola and Raviola (1967) have shown myofibrils which lack thick filaments. The crossbanding patterns also resemble those of skeletal muscle; A, I, H, and M bands, and Z lines are evident in most complete fibrils when sectioned in a proper plane. Corresponding bands of adjacent myofibrils are positioned directly opposite each other in some cases, but this alignment "in register" is never as complete as that typical of skeletal muscle. The absence of fiber alignment keeps the cross-striations from being obvious when myoid cells are viewed with the light microscope, even though many myofibrils are present. This is particularly true for rounded myoid cells.

Not all myoid cells contain fully developed myofibrils. Some cells, even in adult reptiles, resemble some of the earliest stages of development of the skeletal muscle fibers. Scattered primitive Z lines with associated myofilaments occupy the perinuclear cytoplasm of these myoblast-like cells. Accumulations of rough and smooth endoplasmic reticulum, as well as mitochondria, are concentrated peripherally in these immature myoid cells.

Immature and mature myoid cells are often directly apposed to epithelial cells. Desmosomes occasionally connect epithelial and myoid cells. Mature myoid cells may also border on connective tissue spaces;

astructural and cytochemical study of reptilian blood, marrow, and mus should add much interesting information useful for positive entification and phylogenetic comparison of granulocytes.

Plasma cells occur with some regularity in the connective tissue and in the thymic parenchyma. Rough endoplasmic reticulum is abundant throughout their cytoplasm. The cisternae of the rough endoplasmic reticulum are sometimes relatively flat in profile and sometimes markedly dilated by a light, homogeneous material.

Mucous cells and macrophages are few in number. The cytoplasm of mucous cells is filled with flocculent oval inclusions which resemble the mucous droplets of intestinal goblet cells. Phagocytosed material and lipid droplets are found in varying amounts in macrophage cytoplasm.

Thymic Involution

The thymus of young reptiles has distinct cortical and medullary areas. The cortex is distinguished primarily by its very high density of thymocytes, whose relative proportions decrease with increasing age leading to reduced distinction between the cortex and medulla. The weight and size of the organ also decrease with age and blood vessels and connective tissue occupy a progressively larger portion of the gland.

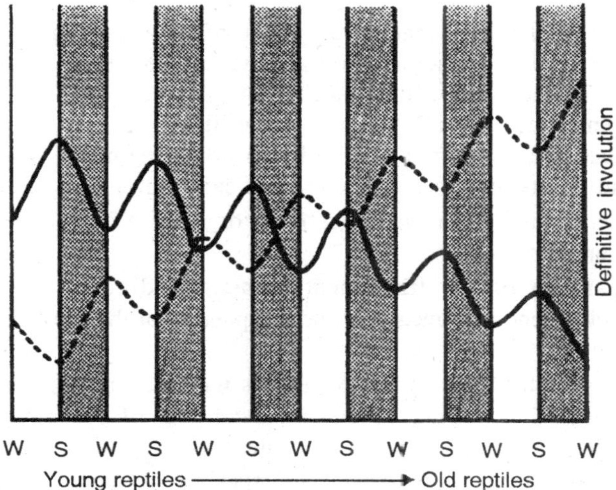

Fig. 7.7. Schematic representation of the periodic seasonal variation in the number of thymocytes (solid line) and the inverse proportion of blood vessels and connective tissue (dotted line) in reptiles. This variation is superimposed on the definitive involution due to age. W, winter; S, spring.

Watney (1882) described three parts in the involuting thymus of a tortoise (*Testudo*): an outer cortical part consisting of flattened cells and intervening reticulum, a medullary portion containing most of the concentric corpuscles, and a zone of tissue between the first two parts which contained a ring of vessels and a large number of thymocytes. The involuted thymus contains hardly any lymphoid cells.

Factors other than age may cause thymic involution in reptiles. Starvation causes reversible changes similar to those described above. Dustin (1909) distinguished between definitive and seasonal involution. The former included involution with age and accidental involution resulting from factors such as disease and starvation. Seasonal involution was considered transitory, and its variation was superimposed on the decrease in size due to definitive involution.

Steroid hormones have a marked effect upon thymic size in mammals and it might be inferred that similar mechanisms exist in reptiles. However, no definitive experiments to determine the presence and extent of hormonal control of the thymus have yet been carried out.

Conclusion

This chapter has summarized the knowledge gained from approximately a century of work on the reptilian thymus. Of necessity, it has been primarily descriptive in nature, rather than analytical. With the relatively recent development of new techniques such as electron microscopy and autoradiography, and improvements in tissue culture, histochemistry, and cytochemistry, there exists the potential for a tremendous expansion of this knowledge. Utilizing the available techniques within the framework of planned experimental procedures, it should be possible to contribute significantly in at least four areas:

1. understanding of the variation of the organ in different species and group of reptiles;
2. expansion of the fundamental understanding of the origin, development and function of the components of the reptilian thymus *per se*;
3. comparison of the thymus of reptiles with that of other vertebrate classes, thus adding to the understanding of the phylogenetic development and relationships of the thymus, and
4. use of the reptilian thymus as an experimental model for study of pathological alterations observed in humans.

Many other questions about the reptilian thymus remain unanswered. Is the reptilian thymus necessary during ontogenetic development of adaptive immunity ? If so, when does this take place ? Does the reptilian thymus participate in the immune response ? Does it produce antibodies ? From which cells do thymocytes originate ? Are thymocytes different from other lymphoid cells ? How do myoid cells originate ? Do myoid cells differentiate within the adult thymus? Do they serve as an antigen against which antibodies are produced under certain conditions? Do hormone-producing cells occur in the thymus ?

Some of these questions have been answered for other animals. Some have not. Complete understanding of thymic function can be approached, however, only through consideration of its varying role in all the classes of animals in which it is found.

INDEX